Tissue Repair and Reconstruction

Series Editors

Andy H. Choi, Carlingford, NSW, Australia

Besim Ben-Nissan, Sydney, NSW, Australia

SpringerBriefs in Tissue Repair and Reconstruction provides a unique perspective and in-depth insights into the latest advances and innovations contributing to improved and better treatments for patients with damaged soft and hard tissues as a result of diseases, trauma, and implantations. The book series consists of volumes that offer biomedical researchers better insights into the advancements of biomaterials science and their translation from the laboratory to a clinical setting. Similarly, the series provides information to surgeons and medical practitioners on novel ideas in biomedical science and engineering on top of disseminating new ideas and know-hows in diagnostics and treatment options for patients from head to toe.

The series will cover a number of key topics:

Fundamental Concepts and Surface Modifications: The topic will provide detailed information on the discovery and advancements of biomaterials surface modification approaches and their use within the human body in a safe manner and without provoking any negative tissue response.

Computational Simulations and Biomechanics: Anatomically accurate computational models are being in all fields of medicine particularly in orthopedics and dentistry to reveal the biomechanical functions and behaviors of bones and joints when damaged, diseased, and in the health state. They also contribute to our understanding during the design and applications of implants and prosthetics subjected to functional loadings and movements.

Surgical Advances and Treatment Options: Discusses how surgical techniques are revolutionized by our deeper understanding into biomaterials science and tissue engineering. The section also focuses on the latest innovations and surgical advancements currently being used to treat patients with damaged tissues.

Post-Operative Treatment and Rehabilitation Engineering: Expands the independence and functionality of the patient after surgery while at the same time reducing the chance of complications such as wound infections and dislocations. Advances in technologies are creating new opportunities in how physiotherapy rehabilitations are delivered.

Necdet Sağlam · Feza Korkusuz · Mesut Şam
Editors

Nano-Biomaterials in Tissue Repair and Regeneration

Clinical Aspect for Hard Tissues Materials

 Springer

Editors
Necdet Sağlam
Nanotechnology and Nanomedicine
Division
Hacettepe University
Ankara, Türkiye

Feza Korkusuz
Department of Sports Medicine
Faculty of Medicine
Hacettepe University
Ankara, Türkiye

Mesut Şam
Department of Biology
Aksaray University
Aksaray, Türkiye

ISSN 2731-9180   ISSN 2731-9199 (electronic)
Tissue Repair and Reconstruction
ISBN 978-981-96-1340-3   ISBN 978-981-96-1341-0 (eBook)
https://doi.org/10.1007/978-981-96-1341-0

This Springer imprint is published by the registered company Springer Nature Singapore Pte Ltd.
The registered company address is: 152 Beach Road, #21-01/04 Gateway East, Singapore 189721,
Singapore

If disposing of this product, please recycle the paper.

Preface

Clinical regeneration of bone and joint cartilage tissues is currently moving to a new dimension. Regenerative medicinal approaches are advancing by innovative research and technological breakthroughs in nanotechnology and new biomaterials. Advanced technology in omic studies as well as artificial intelligence has started to dominate our research journeys. This book aims to provide a comprehensive overview of the latest developments in the research and application of nano-biomaterials for the repair and regeneration of hard tissues.

Regeneration of bones that are common after gunshot wounds, high-energy road traffic, and workplace accidents emerge with complex and open fractures. The incidence of infection has to be prevented and/or treated. Vascularization should be re-established. Fractures in osteoporosis are another challenge in an aging population. Traditional methods of treatment, while effective to some extent, often fall short in terms of long-term outcomes and functionality. This is where nano-biomaterials come into play, offering novel solutions that enhance the efficacy and success of tissue repair and regeneration. Nano-biomaterials, due to their unique properties at the nanoscale, can mimic the natural extracellular matrix, promote cellular activities, and provide mechanical strength to biological structures. Their ability to integrate with biological systems at the molecular level makes them ideal candidates for developing advanced therapeutic strategies.

We explore nanoparticles, nanostructures, and smart piezoelectric materials of various types and their roles in hard tissue repair and regeneration. We furthermore give insights into omic research and artificial intelligence as they interact with biological tissues and revolutionize clinical practices.

Our goal is to provide recent researcher outcomes to clinicians and students with a thorough understanding of future directions of nano-biomaterials in hard tissue repair and regeneration. We hope that this book will inspire further research and innovation, ultimately leading to improved patient outcomes and the advancement of medical science.

Ankara, Türkiye   Necdet Sağlam
Ankara, Türkiye   Feza Korkusuz
Aksaray, Türkiye   Mesut Şam

Contents

Nanoparticles in Tissue Repair 1
Gözde Koşarsoy Ağçeli and Kanika Dulta

Nanostructured Titanium Surfaces in Hard Tissue Repair 27
Eylül Yakar, Boğaç Kılıçarslan, and Cem Bayram

**Smart Piezoelectric Materials for Hard and Cartilage Tissue Repair
and Reconstruction** 59
Sevin Adiguzel, Serap Sezen, and Feray Bakan Misirlioglu

**Artificial Intelligence in Predicting Hard Tissue Regeneration:
Current Situation and Upcoming Perspectives** 73
Nura Brimo, Dilek Çökeliler Serdaroğlu, Halit Muhittin,
Mustafa Kaplan, and Abdulwahab Omira

Nanoparticles in Tissue Repair

Gözde Koşarsoy Ağçeli◉ and Kanika Dulta

Abstract Tissue engineering is a multidisciplinary field of study that offers alternative options to preserve existing tissue or support new tissue formation. One of the main purposes of tissue engineering is to create tissue scaffolds to repair or imitate existing tissues and organs that need remodeling. Nanoparticles, which have the most common use in nanomaterials, are particles ranging in size from 1 to 100 nm. Nanoparticles have contributed to research especially in the field of biomedicine due to their superior properties. The most important advantages of applying particles smaller than 100 nm in biomedical uses are: they show high efficiency surface areas, high stability, low settling rates, and enhanced textural diffusion, which will allow the ligands to be easily attached to the surface. In this book chapter, the role of nanoparticles and composites obtained with nanoparticles in tissue repair will be discussed.

Keywords Tissue engineering · Nanoparticles · Biomedical · Tissue repair · Bone · Cardiac · Neural · Dental · Skin · Tissue repair

1 Introduction

Nanomaterials consist of different materials such as inorganic metals, synthetic or organic polymers, ceramics. These materials can be synthesized physically, chemically, or biologically and have different properties according to the method of synthesis [15].

Nanotechnology has many different uses, and tissue and organ regeneration is among these areas of use (Fig. 1). The use of nanotechnology in the field of tissue and organ regeneration is developing with increasing interest. Nanoparticles, which

G. Koşarsoy Ağçeli (✉)
Faculty of Science, Department of Biology, Hacettepe University, 06800 Beytepe Campus, Ankara, Turkey
e-mail: gozdekosarsoy@gmail.com

K. Dulta
School of Applied and Life Sciences, Uttaranchal University, Dehradun 248007, India

© The Author(s), under exclusive license to Springer Nature Singapore Pte Ltd. 2025
N. Sağlam et al. (eds.), *Nano-Biomaterials in Tissue Repair and Regeneration*, Tissue Repair and Reconstruction, https://doi.org/10.1007/978-981-96-1341-0_1

1

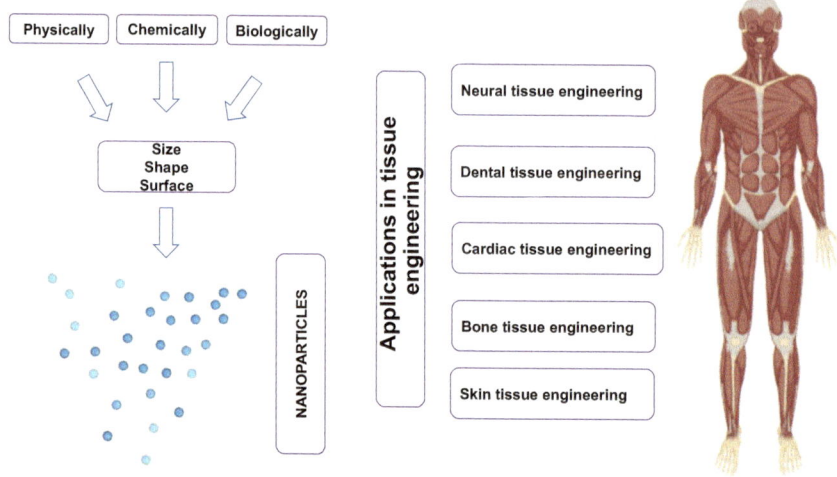

Fig. 1 Various areas of use of nanoparticles in tissue engineering

are a part of nanotechnology, display interesting properties in the development of regenerative medicine [109]. Nanoparticles (NPs) constitute a wide category of materials encompassing particulate substances with dimensions of at least 100 nm [68]. Nanoparticles can be used as vectors for the transport of drugs, growth factors, and genetic material, thanks to their surface area [131]. Nanoparticles such as silver, gold, titanium dioxide, carbon nanomaterials, nucleic acids, liposomes, organic polymers, and iron oxide can be used as materials for medical applications [25, 97, 105, 143, 152].

Superparamagnetic nanoparticles may produce a significant magnetic moment when they are subjected to an external magnetic field because of their nanoscale size, which allow them to attain saturation magnetization. Nonetheless, a particle's size affects its magnetic moment somewhat since bigger volumes result in stronger magnetic attraction. Larger particles that remain in the superparamagnetic region are thus preferred as emphasized by Estelrich et al. [27]. It is crucial to also factor in nanoparticle size concerning biocompatibility. The body decomposes larger particles (with a diameter exceeding 200 nm) through spleen filtration and, ultimately, by phagocytic cells. Smaller particles (with a diameter less than 10 nm) undergo extravasation and renal clearance. Particles falling within the 10–100 nm diameter range are typically the preferred option for intravenous injection since they can navigate through the body's reticuloendothelial system and access small capillaries.

1.1 Nanoparticles in Bone Tissue Engineering

Bone is a multifunctional tissue that not only provides mechanical support and protection, but also facilitates movement in the living body [108].

Biopolymer composites and ceramic materials can be used as scaffolds for bone tissue engineering. However, insufficient mechanical strength of these materials causes limitations in scaffolding. Instead, scaffolds containing composites containing organic and inorganic nanoparticles can be used in bone regeneration and also have advantageous features such as cell structuring, differentiation, and optimal support in scaffolds [57] (Fig. 2).

The next generation of innovative functional materials is expected to lead with nanocomposites or polymer scaffolds/hybrids functionalized with nanoparticles. These materials creatively combine the advantageous properties of the reinforcement and host components. The focus of current research is on creating and designing functionalization of nanocomposite materials in different polymer hybrid systems. This attempts to improve these hybrid scaffolds' toughness, swelling behavior, gel fraction, tensile strength, and tissue regeneration capacities, among other biophysical attributes [70].

As discussed by Vieira and his research team, various types of nanoparticles (NPs) hold promise in the field of bone tissue engineering, particularly in the context of scaffold enhancement and drug delivery. In the domain of nanoparticles (NPs), both inorganic and organic, gold nanoparticles (AuNPs) have been used in scaffolds to support bone regrowth, mostly because of their capacity to induce cell differentiation [59, 130, 149].

Lately, there has been a growing fascination with metal nanoparticles, particularly gold nanoparticles (GNPs), in the realm of bone regeneration. Their popularity has surged significantly, driven by their straightforward manufacturing process, stability,

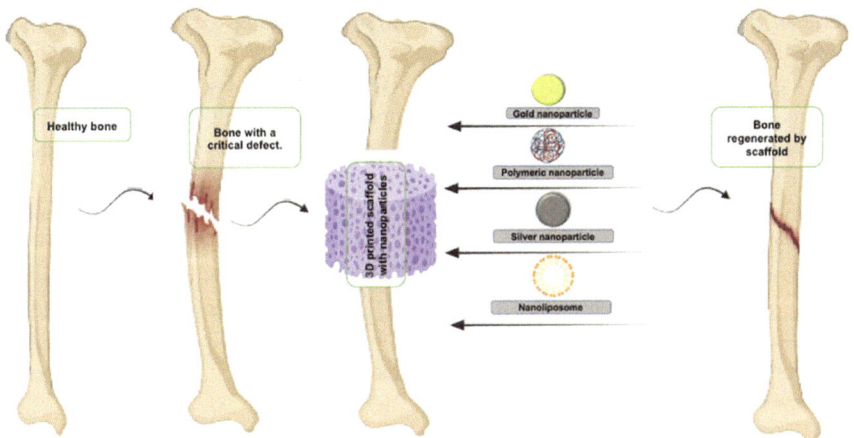

Fig. 2 The role of 3D scaffolds containing different nanoparticles in bone regeneration

adjustable size, extensive surface area, simple functionalization, and versatility [129]. Within the field of bone tissue engineering, gold nanoparticles (GNPs) have the capability to enhance their multifunctionality across various aspects involved in tissue engineering, including scaffolds, cells, and bioactive cues. GNPs can enhance the physicochemical characteristics of scaffolds, rendering them better suited for bone regeneration. Moreover, they promote cellular activities by augmenting adhesion, proliferation, migration, and differentiation [43].

Because HAP (hydroxyapatite) and MNP (magnetic nanoparticles) have been shown to improve bone tissue mineralization when included into different scaffold materials, they are often used in the field of bone tissue engineering. These materials have also found specific applications as bioactive components within hydrogels, further advancing mineralization and fostering osteogenic differentiation. Various matrices have been combined with magnetic nanoparticles to produce magnetic composites. In fact, in animal models, a number of research groups have shown the beneficial effects of mild magnetic fields or pulsed electromagnetic fields on processes including spinal fusion, bone ingrowth into ceramics, and bone fracture repair. Moreover, it has been shown that, both in vitro and in vivo, a high static magnetic field between 0.1 and 10 Tesla affects the alignment of matrix proteins and cells [20, 32, 62, 93].

Cell proliferation and osteogenic differentiation are significantly impacted by the density of Magnetic Nanoparticles (MNPs) in a scaffold, in addition to its mechanical qualities. While elevating the MNP content in scaffolds can enhance the magnetic microenvironment, there exists a critical threshold beyond which cell activity noticeably decreases. Research looking at various MNP concentrations has shown that increased MNP content tends to promote both osteogenic differentiation and cell proliferation. However, optimum cell performance, including proliferation and ALP activity, was reached at 3 and 15% MNP content, respectively, in both PCL and CPC scaffolds. After that, a noticeable drop was seen. These results highlight the possibility of improved cell function with higher MNP content, but they also highlight the need to carefully select the ideal concentration in order to avoid any harmful consequences [12, 55, 120, 121, 139].

In a particular research study, composite materials combining a cyclic copolymer known as COC (Cyclic Olefin Copolymer), often referred to as TOPAS, and Hydroxyapatite (HA) were synthesized for applications in bone repair and regeneration. Within these composites, the HA nanoparticles were evenly distributed within the TOPAS copolymer matrix, leading to enhanced mechanical properties attributed to better dispersion and improved interfacial interactions with the HA chains. Preosteoblasts, which are cells involved in early bone formation, exhibited superior adhesion and proliferation on the TOPAS/HA hybrid materials when compared to TOPAS alone. The study highlighted the enhanced biomedical potential of TOPAS/HA hybrid materials over TOPAS alone, underscoring the advantages stemming from the improved mechanical properties of these materials [2].

Certain metal oxides, especially silver nanoparticles, have demonstrated notable antimicrobial properties and wound-healing potential. Xing et al. conducted an evaluation of the antibacterial effects of nanofibrous scaffolds made from PHBV (poly(3-hydroxybutyrate-co-3-hydroxyvalerate) including Ag. Their findings revealed that PHBV nanofibrous scaffolds infused with silver displayed strong antibacterial characteristics and exhibited excellent in vitro cell compatibility. These results suggest that PHBV nanofibrous scaffolds incorporating AgNPs hold promise for applications in joint arthroplasty and warrant further research and investigation [3, 56, 140].

In addition to AgNPs, various metal oxide nanoparticles have demonstrated their effectiveness as bactericidal agents. Notably, MGONPs (magnesium oxide nanoparticles), and their halogen derivatives with Cl_2 and Br_2, have exhibited robust antibacterial properties. To evaluate the efficacy of these NPs, a study conducted by Stoimenov et al. focused on three bacterial groups: *E. coli, B. megaterium,* and *B.subtilis.* The results demonstrated that nanoparticulate formulations could eliminate populations of *E.coli* and *B. megaterium* within approximately 20 min. Moreover, there is growing interest in the use of SeNPs, as they have been shown to have distinct antibacterial mechanisms compared to the aforementioned NPs, which rely on the generation of reactive oxygen species (ROS) for their bactericidal effects [123, 134].

Another interesting option for bone tissue engineering applications is nano-sized bioactive glasses, or nBGs. Comparing dense nBGs to mesoporous bioactive glass nanosphere composites, Covarrubias et al. found that the addition of dense nBGs to a chitosan–gelatin polymer mix increased alkaline phosphatase activity. This implies that nBGs may improve the uses of bone tissue engineering [14].

1.2   Nanoparticles in Skin Engineering

The skin serves as the initial defense mechanism of the human body. It is subjected to outside stimuli, thereby playing a vital role in safeguarding against injuries and harm [77, 113]. One consequence of the skin being exposed to injuries is aging, and another is wounding. Due to this, the restoration of tissues and the revitalization of skin have become areas of focus within clinical practice [9].

The advent of nanotechnology has brought about a revolution in the realm of wound dressings. It has led to the creation of different materials and medication delivery methods based on nanotechnology which are developed to address challenging-to-heal wounds (Fig. 3). Chronic wounds present a formidable challenge, requiring highly effective therapies to combat the complexities of an impaired healing process. While numerous novel approaches have fallen short in delivering precise healing outcomes, this has made it possible for a number of nanotechnology-based wound-healing techniques to surface [115].

Nanomaterials offer distinct advantages over other wound dressings, including their flexibility, low cytotoxicity, excellent biocompatibility, drug delivery capabilities, and their ability to exhibit a wide range of physicochemical properties, equipping

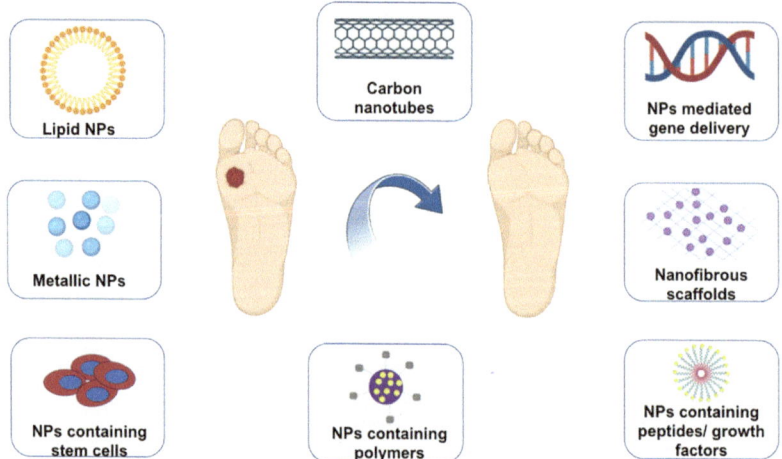

Fig. 3 Different nanotechnological materials in wound treatment

them with unique attributes. Their nanoscale dimensions facilitate enhanced penetration to the injury site and increase the likelihood of effective interactions with biological targets. Additionally, nanoparticles (NPs) have the capacity for controlled and sustained release of therapeutic agents, leading to expedited wound-healing processes [48, 61, 96].

The potential of various metallic nanoparticles, polymeric nanoparticles, peptide-loaded nanostructures, and carbon-based nanomaterials in wound healing has been thoroughly investigated. These materials offer remarkable physical, chemical, and biological properties, including ease of fabrication, biodegradability, and high biocompatibility, making them promising candidates for applications in wound healing [34, 65].

Gold nanoparticles (AuNPs) exhibit superior biocompatibility compared to other metallic nanoparticles, making them a promising option for wound-healing applications. It's worth noting that AuNPs, both alone and in combination with other drugs, have been extensively explored for their wound-healing potential. AuNPs synthesized using an aqueous extract of Citrullus lanatus rind have demonstrated proteasome inhibitory activity, antibacterial properties, and antioxidant potential, making them viable candidates for wound healing. Incorporating pharmaceutical intermediate-capped AuNPs into electrospun scaffolds has shown promise in treating full-thickness wounds contaminated with germs resistant to multiple drugs (MDR). By modifying membrane potential and blocking the ATP synthase enzyme, AuNPs' antibacterial method of action causes disruptions in cellular energy metabolism and eventually results in cell death [38, 144].

Zinc oxide (ZnO) nanoparticles' intrinsic antibacterial qualities make them ideal for a range of hydrogel-based wound treatments. In a study, wound dressings made of cotton and infused with silver nanoparticles (AgNPs), zinc oxide nanoparticles (ZnO NPs), and a combination of both (Ag/ZnO NPs) exhibited strong antibacterial activity, highlighting their potential for effective wound care [51]. Zinc oxide nanoparticles (ZnO NPs), derived from an aqueous leaf extract of the *Barleria gibsonii* plant, have demonstrated remarkable wound-healing properties in rat models. These nanoparticles have proven to be an effective and superior topical antimicrobial treatment for burn wounds [118].

Wound dressing materials frequently rely on polymeric nanostructures, which can be composed of either synthetic or natural polymers. The synthetic polymers poly(lactic-co-glycolic acid) (PLGA), poly(ethylene glycol) (PEG), poly(lactic acid) (PLA), and polycaprolactone (PCL) are often used for creating biomaterials in wound care applications. Additionally, hyaluronic acids, chitosan, cellulose, and alginates—all naturally occurring biodegradable polymers—have been shown to be important in the healing of wounds. In one particular research, curcumin-loaded electrospun nanofibers made from a combination of PLA and hyperbranched polyglycerol shown significant wound-healing ability. Within 36 h, the wound may be fully covered by the curcumin-loaded electrospun nanofibers, according to the results of in vitro scratch tests [100].

Human skin exhibits permeability to 6 nm particles when they are suspended in an aqueous buffer. However, the penetration of 15 nm particles is more feasible when using toluene as the medium. Additionally, through the utilization of a lipid-based permeation assay that closely mimics the lipid composition of the human stratum corneum (SC), it has been observed that 100 nm lipid nanoparticles can traverse the skin barrier. Nevertheless, it's worth noting that the concentrations of nanoparticles and released drugs in the dermal matrix (DM) may still remain considerably low [26, 86].

At present, cosmetic products, including sunscreens, hair products, and skin creams, incorporate nanoformulations. While the use of nanomaterials in cosmetics dates back three decades to the introduction of liposomes, ongoing research trends continue to focus on the development of innovative devices. This includes the exploration of diverse materials and techniques, with recent applications extending to areas like wound dressings [8, 67].

Metal-based nanoparticles (such as Au, Ag, CuO, and ZnO) are commonly used in nanocosmetics owing to their recognized antibacterial properties and their ability to promote wound healing. Gold nanoparticles, in particular, have made their way into a variety of cosmetic products, including moisturizers, sunscreens, eye creams, and lip balms, primarily for their unique properties. Additionally, numerous spas and beauty professionals offer facial masks and treatments incorporating gold nanoparticles to rejuvenate the skin [79].

1.3 Nanoparticles in Neural Tissue Engineering

Neurons in the peripheral nervous system are essential for both receiving and sending electrical impulses to and from the central nervous system. The creation of methods to provide medicinal compounds to these peripheral neurons holds great promise for addressing various clinical conditions, including neuropathic pain, neuropathies, nerve injuries, and regeneration [74, 117, 136]. However, it has proved to be a difficult undertaking to distribute medication molecules to peripheral neurons. This challenge primarily stems from the constrained peripheral nerve distribution according to neuroanatomy and the existence of the strong blood–nerve barrier, which provides protection for the nerves' endoneurial milieu. The systemic administration of drugs can result in undesirable off-target effects due to high drug concentrations throughout the body, while local delivery methods, often employed in regional and nerve block anesthesia, have limited applicability [28, 75, 116, 125].

To address the numerous challenges associated with drug delivery to neural tissues, researchers have turned to the use of nanoparticles as a targeted approach. Attempts to administer customized nanoparticles, varying in size and surface modifications, to neural tissues have yielded diverse outcomes. The blood–nerve barrier nevertheless poses a major obstacle to the entrance of nanoparticles, despite their tiny size allowing them to passively pass through most cellular barriers. There has been modest success in allowing partial penetration of nanoparticles into the nerve tissue by local injection of nanoparticles near peripheral nerves. Less is known about their exact distribution inside cell bodies and axons, however. Moreover, it might be technically difficult to directly apply nanoparticles near nerves and requires knowledge of the neuroanatomical position of the nerve. Given these drawbacks, a different method of reaching peripheral axons and, by extension, the neuronal cell bodies might be via the skin. A huge network of axonal terminals, which stand in for the branching ends of peripheral nerves, delicately innervates the skin [22, 35, 40, 110, 124].

The breakthroughs in nanotechnology, which empower the manipulation of matter at atomic and molecular scales, have ushered in remarkable techniques and approaches for the synthesis and engineering of innovative biomaterials. These nanomaterials, characterized by their versatile functionality and adjustable biological attributes, have emerged as promising contenders for neural repair and regeneration. Their efficiency becomes evident in their ability to actively participate in the modification of the microenvironment of nervous tissue, regulate neural responses, and facilitate precise and selective drug and biomolecule delivery strategies [39, 45, 50, 78, 92, 153].

The integration of growth factors (GFs) into nanomaterials and nanodevices, whether through physical or chemical loading methods, stands as a versatile and effective approach used in the development of neuroregenerative and neuroprotective strategies aimed at peripheral and central neural therapies [29, 111].

The potential of Dextrin–GF conjugates as long-term targeted delivery vehicles for bioactive proteins essential for brain regeneration has been investigated in recent research. These conjugates have demonstrated the ability to enhance and extend the proliferation, specific differentiation of nerve cells, and the suppression of apoptosis in mouse-derived neural stem cells (mNSCs). When exposed to amylase, these biopolymer-protein formulations—which are produced by conjugating succinylated dextrin with basic fibroblast growth factor (bFGF) and epidermal growth factor (EGF)—show persistent growth factor release at concentrations that are comparable to those found naturally in cerebrospinal fluid. Dextrin–GF conjugates increase nestin protein expression (linked to neural progenitors) and control the expression of markers linked to the development of neurons, astrocytes, and oligodendrocyte cells in mNSC cultures [31].

Zamproni and associates looked at spherical micro- and nanoparticles made of poly(lactic-co-glycolic acid) (PLGA) as possible carriers for stromal cell-derived factor 1 (SDF-1) encapsulation and controlled release. A chemokine called SDF-1 is in charge of promoting neuron migration throughout the development of the embryo. With mean particle sizes of 4.83 ± 0.33 μm and 167.9 ± 0.38 nm, respectively, the methods they created demonstrated remarkable encapsulation effectiveness for SDF-1, exceeding 84%. Additionally, these technologies made it possible for the protein to be released effectively and continuously; the PLGA nanoparticles showed very lengthy release profiles. By releasing SDF-1 at physiologically comparable amounts, both polymeric formulations created the ideal milieu for cellular chemotactic responses. Notably, the PLGA/SDF-1 system especially showed neurogenic potential when it was in the form of nanoparticles. This was shown by the notable healing of traumatic mouse brain lesions, which was made possible by increased neuroblast recruitment to the injury site [148].

Researchers investigated the synergistic effects of simultaneous release of chondroitinase ABC (ChABC) and SDF chemokine in another remarkable study. Proteus vulgaris-derived enzyme ChABC is essential for the hydrolysis of chondroitin sulfate proteoglycans, which are present in glial scarring and block myelin proteins in the nervous system. The goal of this combination strategy was to increase brain tissue's capacity for regeneration. Local injection of methylcellulose hydrogels cross linked with SDF-loaded PLGA nanoparticles or ChABC-specific peptides and proteins in the thoracic region improved persistent locomotor activity in a rat model for a period of up to 8 weeks. The methylcellulose hydrogels loaded with both ChABC and SDF, as well as those supplied with chemokines alone, showed a decrease in the amounts of certain glycosaminoglycan chains in glial scarring. Furthermore, these formulations promoted the improved migration of brain precursor cells found in the body [95].

Iron oxide nanoparticles (IONPs) possess the capability to breach the blood–brain barrier (BBB), rendering them valuable for numerous applications within the central nervous system (CNS). These applications encompass targeting amyloid beta (Aβ) in cerebral arteries, suppressing microglial cells, facilitating drug delivery, and enhancing contrast in magnetic resonance imaging (MRI) [21, 37, 102].

Iron oxides, more especially magnetite (Fe_3O_4) and maghemite (γ-Fe_2O_3), are the main magnetic minerals used in physiological applications. These two materials share a closely related structure, resulting in similar characteristics. However, they exhibit slight variations in their saturation magnetization values, with magnetite ranging from 92 to 100 emu/g and maghemite from 60 to 80 emu/g. Additionally, their superparamagnetic diameters differ, measuring 25 nm and maghemite measuring 30 nm in diameter [27]. However, because of their notable toxicity, several other magnetic minerals, such as cobalt and chromium, are often avoided in biological applications. To enhance magnetic susceptibility, iron-based metal oxides such as $CoFe_2O_4$, $NiFe_2O_4$, and $MnFe_2O_4$ can serve as the core of nanoparticles. However, it is imperative to apply impermeable coatings to prevent the release of these toxic metals into the surrounding tissue, as noted by McBain et al. [82] and Kudr et al. [63].

In vitro tests were conducted to evaluate the effects of different magnetic liposomes and polymer-encapsulated magnetic nanoparticles on neurons as part of the first studies into the use of superparamagnetic iron oxide nanoparticles (SPIONs) for medication delivery inside the nervous system.

The present study proposed by Kim et al. included the synthesis of magnetic liposomes by the integration of superparamagnetic iron oxide nanoparticles (SPIONs) coated with oleic acid, along with a polyethylene glycol (PEG) chain that was coupled to a phospholipid tail. Subsequently, the magnetic liposomes were assessed by in vitro research using PC12 cells, a widely used rat-derived cell line for investigating neural development and proliferation. In the course of this investigation, PC12 cells were subjected to various concentrations of magnetic liposomes for a period of up to 5 days. The findings of the study revealed that the magnetic liposomes demonstrated a correlation between the dosage administered and the stimulation of neurite outgrowth. This effect was more pronounced when the magnetic liposomes were used in conjunction with exogenous nerve growth factor (NGF). In addition, the introduction of liposomes and NGF resulted in the elevation of phosphorylated ERK1/2, a protein known to be involved in the NGF signaling pathway. Moreover, this intervention led to heightened expression of β-tubulin and integrin $\beta1$ in PC12 cells [54].

Zuidema et al. used an innovative strategy to direct the development of primary rat dorsal root ganglion (DRG) neurites. This included creating a gradient of nerve growth factor (NGF) by means of controlled release from magnetic nanoparticles (MNPs). The MNPs were coated with poly-L-lactic acid (PLLA) and then infused with a concentration of 3 ng/mL of nerve growth factor (NGF). The examination of release kinetics demonstrated an initial burst release of 1.1 ng of NGF during the first a day period, followed by a more prolonged release of nanograms of NGF over a duration of 7 days. In order to promote targeted neurite outgrowth on a specific side of the dorsal root ganglion (DRG), the researchers incorporated magnetite nanoparticles (MNPs) loaded with nerve growth factor (NGF) into aligned poly-L-lactic acid (PLLA) microfibers. This resulted in the development of a hybrid biomaterial system that effectively affected the extension of neurites. The NGF gradient was established with the strategic placement of a neodymium magnet at a distance of 5 mm from the dorsal root ganglion (DRG) body. This arrangement effectively drew the NGF-loaded magnetic nanoparticles (MNPs) toward the desired location. The neurites that emerged stretched bidirectionally along the fibers. However, it is noteworthy that they displayed much longer length on the side that was in closer proximity to the NGF-releasing particles [155].

One of the main challenges in the field of neural regeneration, namely, in the peripheral and central nervous systems, is to the efficient and long-lasting administration of bioactive substances. The toxicity of conventional systemic delivery techniques is a concern, and the post-injection stability of growth hormones is constrained. Nevertheless, a potential resolution is the coupling of growth factors with superparamagnetic iron oxide nanoparticles (SPIONs), a strategy that has been shown to prolong their half-life while maintaining their biological functionality. In a prominent scientific endeavor, a team of researchers successfully attained a remarkable 70% efficacy in the adsorption of brain-derived neurotrophic factor (BDNF) onto magnetite nanoparticles with a diameter of 60 nm. In their study, the researchers used an in vitro representation of the blood–brain barrier (BBB) to illustrate that the application of an external magnet might enhance the transportation of adsorbed brain-derived neurotrophic factor (BDNF) via magnetite nanoparticles (MNPs). Specifically, the results indicated that the magnet-induced transportation facilitated 3.5 times more movement of adsorbed BDNF compared to the un-adsorbed BDNF while crossing the in vitro BBB model. Significantly, the transportation of BDNF-MNPs did not impair the integrity of the blood–brain barrier (BBB) model, thereby underscoring its potential as a viable strategy for the efficient and targeted administration of bioactive substances to the nervous system [101].

In a study conducted by Kong et al., it was demonstrated that the study provided evidence that particles with a diameter of 124 nm, coated with polystyrene and magnetite, were seen which, on their own, do not traverse the blood–brain barrier (BBB). However, when subjected to the impact of an outside magnet positioned near the mouse's head, these particles were guided across the BBB. While the majority of the particles were eliminated from circulation and accumulated in the liver and spleen, an intriguing finding revealed that approximately 30% of these particles remained localized in the brain even after a 48-h period [60].

Das et al. developed a gold nanocomposite using silk as the foundation material, with the intention of enhancing the attachment of Schwann-like cells to improve the process of peripheral nerve regeneration [16].

The size and shape of gold nanoparticles (AuNPs) can have a significant impact on their potential to induce neural differentiation. Research indicates that 30 nm AuNPs have shown remarkable efficacy in promoting the differentiation of embryonic stem cells (ESCs) into dopaminergic neurons, while also exhibiting excellent biocompatibility [135].

Despite possessing various positive qualities, nanoparticles give rise to health risks as a result of their diminutive size and chemical compositions. Scientists have shown a strong interest in examining the potential adverse impacts of nanoparticles on neuronal function. Recent research findings have shown that the use of diluted amounts of silver nanoparticles (AgNPs) might result in adverse effects on brain health. Furthermore, previous studies have shown that the administration of silica nanoparticles has a detrimental effect on the functionality of mitochondria throughout the process of neural development. In a further investigation, it was noted that the introduction of polyamidoamine (PAMAM) dendrimers including a range of surface functional groups resulted in cytotoxic consequences on the process of neuronal differentiation in human neural progenitor cells. The nanoparticles, upon examination under controlled laboratory circumstances, were seen to facilitate harm to neurons, initiate the process of neurodegeneration, lead to toxicity in neuronal cells, and demonstrate neurotoxic properties. Furthermore, alongside in vitro models, nanoparticles have been subjected to experimentation in animal models, resulting in the induction of neuronal injury, degeneration, toxicity, cell death, and the manifestation of deleterious effects on the blood–brain barrier (BBB) [42, 150].

The findings of the study indicate that silica nanoparticles do not facilitate differentiation and instead have a negative effect on the process of stem cell differentiation. The investigation conducted by Ducray et al. examined the internalization of silica nanoparticles by SH-SY5Y cells, elucidating a decrease in neural development. In contrast, several investigations have offered an alternative viewpoint about silica nanoparticles. An experiment conducted by Kouki Fujioka and colleagues included the synthesis of silica nanoparticles of different diameters. The results of this study demonstrated that all of these particles exhibited an upregulation in the expression of Nestin and N-FH. In addition, the use of nanoparticles measuring 30 and 44 nm at a concentration of 0.1 mg/mL resulted in an elevation in GFAP expression, indicating the facilitation of self-renewal and neural differentiation of human neural stem cells (hNSCs). Interestingly, a modest decline in HMGA1 was seen after the administration of 30 nm silica nanoparticles, suggesting a potential decrease in neurogenesis of human neural stem cells (hNSCs) [33].

1.4 Nanoparticles in Dental Engineering

Teeth are located in the mouth and consist of different components such as dentin, enamel, cementum, pulp, and the periodontal ligament. Their primary role is to slice and grind food, making it more manageable to swallow and digest [13].

The use of nanoparticles in the area of dentistry has promise for augmenting therapies at the atomic and molecular scales. In general, this technological approach entails the use of nanoparticles with dimensions ranging from 10 to 100 nm in order to augment the properties of conventional materials by the integration of functional groups. The primary objective is to develop approaches for disease prevention, diagnosis, and treatment [92]. The field of dentistry has seen significant advancements in the use of nanoparticles across a range of applications. These include dental implants, preventative and antibacterial nano-dentistry, restorative dentistry, medication delivery, oral cancer, tissue regeneration, periodontics, dentin hypersensitivity, and other related fields [87, 90, 103].

Silver nanoparticles are among the most commonly utilized NPs, found in various product forms, and they are closely followed by carbon and ion oxides, such as TiO_2. These nanoparticles enhance product quality by incorporating numerous functional groups. Consequently, nanoproducts find extensive applications across diverse industrial sectors, in the field of medicine, and notably within the realm of dentistry [53, 83, 141].

Dental implants, which consist of biocompatible substances like hydroxyapatite and titanium, are surgically placed inside the alveolar bone alongside a prosthetic tooth, making them suitable for installation. Nevertheless, these implants primarily develop a periodontium that exhibits distinct characteristics compared to the composition of the native tissue. When the alveolar bone comes into direct contact with the dental implant, it can potentially lead to dental ankyloses. In order to tackle this matter, nanoporous anodic alumina (NAA), porous silicon (pSi), and Titania nanotubes (TNTs) are used in the advancement of drug-releasing implants via the electrochemical anodization technique applied to silicon, titanium, and aluminum [66, 94, 98].

Dental caries, a globally recognized issue that impacts people of all age groups, presents an opportunity for NPs (nanoparticles) to serve as effective tools for reversing cavities caused by bacterial culturation and demineralization. Typically, conventional treatments involve the application of fluoride agents to affected cavities to facilitate tooth remineralization and deter bacterial activity. When administered at the nanoscale, these agents can create more discreet treatment systems that are minimally noticeable to patients and can be precisely targeted to specific areas [47, 87].

The use of calcium fluoride (CaF2) nanoparticles in bioadhesive films has been investigated by researchers as a way to overcome the drawbacks of traditional mouth rinses and toothpaste, which often have a transient effect on oral health. By prolonging the fluoride agents' contact time in the impacted regions, this novel strategy seeks to effectively inhibit the development of bacteria and the creation of biofilms [36].

The feasibility of loading fluoride into chitosan nanoparticles has been explored as a potential avenue for oral drug delivery, possibly through sprays or mouthwash solutions. Research findings have indicated that even minute quantities of fluoride within these nanoparticles can offer a sustained release within the oral environment [91].

Periodontal disease is another common dental condition that is brought on by bacterial infections and the immune system's reaction, which releases toxins. In the chronic stage, the disease degrades bone and periodontal fibers due to immune cells being activated and releasing pro-inflammatory cytokines and reactive oxygen species. As a consequence, taking oral drugs like antibiotics and anti-inflammatories together with invasive surgical treatments is often necessary, which may be uncomfortable and have unintended side effects. Treatment with targeted medication delivery by nanoparticles is a promising and less intrusive method. In order to efficiently attack periodontal pathogenic bacteria, Backlund et al. have produced silica nanoparticles that have the potential to emit exogenous nitric oxide [5, 7].

Biotechnology and medicine both heavily rely on iron oxide (FeOx) nanoparticles. Because they are non-toxic to people and biocompatible, magnetite and maghemite are the two most often used varieties of iron oxide nanoparticles in biomedical research. Iron oxide is also quite biodegradable, which makes it a good choice for in vivo applications. Superparamagnetic iron oxide-based nanoparticles are the most often used in the field of medical research. Antibiotic-resistant bacterial biofilms on dental implants are particularly difficult to eradicate because of their barrier of exopolymers, which encloses microbes in an unbreakable matrix and makes them immune and drug resistant. Currently, nanoparticulate materials are used to fight microbial diseases, and iron oxide nanoparticles are often utilized to remove biofilms from dental implants [68, 88, 112].

Fluconazole (FLZ) stands as a primary active component in the treatment of oral candidiasis. While FLZ effectively interacts with various medications when applied through conventional treatment methods like oral gels, mouth paints, and rinses, its duration of efficacy within the mouth is limited. Consequently, the use of adhesive nanoparticles for delivering this agent offers a promising solution. To explore this potential, a study developed mucoadhesive eudragit (EUD) nanoparticles loaded with FLZ and coated with chitosan. The resultant nanoparticle exhibited no cytotoxic effects and demonstrated remarkable stability. Furthermore, findings from ex vivo and in vivo tests on rabbits presented this innovative approach as an attractive option for reducing the overall drug dosage and minimizing side effects, thereby enhancing treatment alternatives for oral candidiasis [18].

Research leveraging nanotechnology to enhance dental treatments extends beyond addressing oral disorders. Dental implants, which are prone to failure as a result of bacterial issues, can also benefit from the properties of nanoparticles (NPs). With a growing number of dental implants being placed in patients worldwide, a trend expected to yield investments of approximately 4 billion dollars by 2022, studies to mitigate implant failure are on the rise. Coating dental implants with nanoparticles represents a promising approach. This technique allows for selective areas of the implant to remain uncoated, promoting the formation of osteoblasts, ensuring controlled distribution of essential elements, and harnessing the therapeutic effects of nanoparticles for the prevention of peri-implantitis [4, 138, 147].

Calciophosphatic particles are commonly incorporated into bone healing sites, typically in the form of filling nanopowders or within hydrogel carriers, often in combination with other components. This inclusion serves the purpose of facilitating the process of remineralization [69]. Specifically, nanostructured hydroxyapatite (nano-HA) and nano-CaP nanomaterials have garnered significant interest in the last decade. Nano-HA, in particular, has exhibited exceptional biological properties when compared to conventional hydroxyapatite (HA). Moreover, nano-HA has demonstrated notable biocompatibility and bioactivity in relation to bone constituents, likely owing to its chemical composition and mineral structure's resemblance to that of bone tissue [6, 49, 127, 133].

Oral biofilms play a crucial role in the development of dental caries, with *Streptococcus mutans* being one of the ten key etiological agents in the formation of dental caries. Additionally, biofilms are also significant contributors to various periodontal diseases. Dental implants serve as replacements for missing teeth, but their success can be compromised by biofilm-related issues. One promising solution comes in the form of the graphene/zinc oxide nanocomposite (GZNC), which has the potential to combat biofilms caused by *Streptococcus mutans*, thereby addressing a common concern in implant failure [64, 85].

Ceramic materials' intrinsic chemical and physical capabilities are successfully combined in nano-zirconia-alumina compounds. A small fraction of tetragonal zirconia oxide nanoparticles are incorporated in an aluminum oxide matrix inside these nanoparticles. The required qualities of hardness and durability, which are of particular significance in the area of dentistry, are preserved by this special mixture.

When compared to conventional ceramic materials, alumina/zirconia nanocomposites show higher efficacy as novel implant materials. Interestingly, zirconia oxide nanoparticles are useful as a polishing agent in dental operations because they show anti-biofilm action against certain bacteria, such as *E. faecalis* [10, 19, 99].

Hydroxyapatite nanoparticles (NPs) have found extensive applications in both the medical and dental fields. Their composition closely resembles that of teeth and bone, rendering them a biocompatible material well suited for various physiological processes. Nano-sized hydroxyapatite (HAp) particles readily penetrate dental tubules, which serve the crucial role of sealing these openings, preventing the exposure of nerves to external stimuli. Consequently, HAp aids in reducing dental hypersensitivity. Nanoparticles composed of hydroxyapatite (HAp) have an increased surface area, hence augmenting their ability to form strong interactions with proteins, as well as bacterial and plaque pieces. Due to their increased biological activity and reactivity, they possess the capability to efficiently bond to dentin apatite and dental enamel [11, 52].

Nanoparticles of titanium dioxide (TiO_2) are widely used in the dentistry and medical domains. Dental implant implantation may cause complicated antigen/antibody type 1 and type IV responses, which can set up allergic reactions. Furthermore, the attachment of microorganisms to Titania implants has a substantial influence on the process of dental healing and may have enduring consequences for the implants' efficacy [41, 83].

Chitosan nanoparticles (NPs) exhibit a bactericidal effect and facilitate the release of fluoride from dental materials, offering a dual benefit in the prevention of dental caries. In a study conducted by Kumar et al., it was illustrated that incorporating 10 wt.% of chitosan NPs into glass ionomer cement (GIC) not only improved the material's resistance but also enhanced its fluoride release properties [114].

1.5 Nanoparticles in Cardiac Tissue Engineering

Cardiovascular diseases (CVD), which account for one in every four deaths and stand as the primary cause of the phenomenon of mortality is seen in both male and female individuals, encompass many illnesses that impact the heart and blood vessels. Hence, CVD is frequently termed the "silent epidemic" because a significant number of individuals remain unaware of their condition [132].

Cardiovascular nanomedicine (CVN) is a discipline that aims to address the shortcomings of existing cardiovascular disease (CVD) therapies by developing novel approaches. The first nanosystems in CVN were created with the intention of enhancing the bioavailability, stability, and safety of already available medications. When compared to traditional materials, nanomaterials have unique physicochemical characteristics which include a higher surface area-to-volume ratio and higher surface energy, which affect biological interactions and protein adhesion. Therefore, the regulated and targeted administration of diverse functional components with the goal of treating problems linked to lipid metabolism and other CVD-associated illnesses is made possible by nanotechnology, which provides a safe and dynamic platform [106, 107].

In the ongoing effort to combat cardiovascular disease (CVD), the use of designer nanoparticles with targeting ligands has emerged as a promising approach for achieving accurate medication delivery to regions inside plaque and the heart that are specifically targeted. The primary objective of these nanocarriers is to facilitate the targeted delivery of medications to specified therapeutic areas, hence reducing the potential adverse effects on healthy tissues. The aim is to use nanotechnology-based approaches in order to address the difficulties associated with traditional systemic drug delivery methods, such as instability, low bioavailability, poor solubility, restricted absorption, and adverse side effects. Insufficient delivery mechanisms have hindered the successful use of DNA vectors, microRNAs (miRNAs), and stem cells for the purpose of vascular tissue repair and regeneration, alongside pharmaceutical medications. Historically, the focus of cardiovascular nanomedicine research has been on the development of nanoscale carriers that include several functional components, including a nanoparticle core, therapeutic payload, and targeting element. The objective of this approach is to improve the transport of these carriers to the heart and circulatory system, therefore overcoming biological barriers. In recent times, there has been an expansion in the field of cardiovascular nanomedicine. This expansion has moved beyond the development of specialized nanocarriers that target particular anatomical locations. It now includes the incorporation of biosensors, actuators, and devices that can be easily incorporated into different phases of medical treatment. The primary objective of this extension is to optimize the assessment, intervention, prophylaxis, and control of post-operative pain [128, 137, 142].

Researchers have been increasingly intrigued by the potential of nanotechnology-assisted stem cell therapy in the treatment of cardiac diseases. This heightened interest stems from the remarkable characteristics of nanoscale materials, their adaptability for modification and functionalization, and the ability to fine-tune their properties. Among the frequently utilized nanomaterials are metallic nanoparticles, polymeric nanoparticles, liposomes, as well as carbon-based and graphene nanoparticles [15, 81, 119, 122, 145, 146]. Numerous studies have explored the use of nanomaterials in the context of cardiac stem cell therapy, often in conjunction with various types of stem cells.

In contemporary times, a range of conductive AuNPs (gold nanoparticles), including nanowires, nanorods, and nanotubes, have been widely used in the domain of cardiac tissue engineering. The primary reason for this phenomenon may be attributed to the exact control of their shape, optical characteristics, and distinctive surface chemistry [139]. Researchers used the remarkable photothermal heat conversion skills shown by gold nanomaterials to exploit the near-infrared heating qualities of gold nanorods (AuNRs). This approach was employed to develop nanocomposite scaffolds and manufacture cardiac patches, eliminating the need for conventional stitching methods. In the present study, gold nanorods (AuNRs) were integrated into electrospun fibers composed of albumin. The fiber scaffold experienced molecular structural changes upon exposure to near-infrared laser irradiation, resulting in its ability to attach to the myocardial infarction (MI) region in a non-invasive manner. In addition, the utilization of conductive AuNM (gold nanomaterials) has been observed not only in the augmentation of electrical conductivity characteristics in electroconductive textiles (ECTs), but also in their application as electrodes in the construction of intricate electronic devices. These devices are designed for the purpose of detecting electrical signals originating from electroconductive textiles and providing targeted electrical stimulation as required [30].

The progress in nanotechnology has made it feasible to create 3D structures capable of nurturing stem cell growth and promoting their differentiation into functional myocardium. Nanofiber-based scaffolds, renowned for their robust tensile strength and ample surface area, facilitate efficient nutrient and waste exchange with the extracellular environment [1].

The integration of gold into scaffolds has been shown to enhance the functioning of the system. The incorporation of gold nanoparticles and nanowires into these scaffolds has been extensively implemented. The findings from these research have shown that the upregulation of essential proteins such as troponin I, alpha-sarcomeric actin, and connexin-43, which are involved in vital cellular chemical and electrical communication processes, results in enhanced and coordinated contraction of myocytes. The synchronization of cardiac tissue grafting plays a crucial role in ensuring its successful integration. Moreover, the augmentation of scaffolds with growth factors, such as insulin-like growth factor 1 (IGF-1) and vascular endothelial growth factor (VEGF), has shown the ability to improve the development and functionality of cardiomyocytes [17, 23].

Experiments have shown the utilization of nano-sized polyisopropylacrylamide (PNIPAAm) materials combined with hyaluronic acid (HA) as a composite material with finely tuned pore diameters. When combined with stem cells, this approach was investigated for its therapeutic potential in treating myocardial infarction (MI) in both rat and pig models [24].

The enhancement of solubility and biocompatibility in carbon nanomaterials is often achieved through the commonly employed method of functional modification, concurrently diminishing their cytotoxic effects. Currently, carbon nanomaterials are being incorporated into composite conductive biomaterials within the realm of cardiac tissue engineering research. Both graphene and CNT (carbon nanotubes) have demonstrated their capability to notably enhance electrical conduction, cardiac cell adhesion, proliferation, and maturation [73, 80, 84].

In 2013, Kim et al. conducted a study to assess the biocompatibility and cellular responses of cardiomyocytes using graphene films. The findings from their research revealed that graphene exhibited outstanding biocompatibility and had the ability to enhance the adhesion, viability, and contractility of mature cardiomyocytes [58].

Through comprehensive research on the attributes of fullerene (C_{60}) nanoparticles, such as their electrical conductivity and resistance to oxidative stress, an injectable hydrogel was developed. The hydrogel in question is a composite material that integrates fullerene conductive nanomaterial with alginic acid hydrogel. This composite hydrogel is further enhanced by the inclusion of BADSCs, which are used for the purpose of repairing myocardial infarction (MI). The results of these trials indicate that the hydrogel treated with fullerene efficiently improves the retention of stem cells after transplantation, consequently assisting in the healing of myocardial infarction [44].

In a study proposed by Qazi et al. in 2014, an experiment was conducted where they coupled PANi with camphorsulfonic acid and polyglycerol-sebacic acid in different proportions. This combination resulted in the formation of a conductive complex that exhibited remarkable electrical conductivity, mechanical strength, and pH buffering properties. The aforementioned intricate substance has a notable capacity to support the development and proliferation of cardiomyocytes, so establishing itself as a very promising nanomaterial with prospective applications in the field of cardiac tissue engineering [104].

According to available reports, the administration of PLA nanoparticles containing beraprost, a prostacyclin analog, and monomethoxy poly(ethylene glycol)-poly(lactide) block copolymer via intravenous route has been found to offer protection against pulmonary arterial remodeling and right ventricular hypertrophy induced by monocrotaline (MCT). In addition, the injection of beraprost-nanoparticles on a weekly basis has shown efficacy in attenuating the pulmonary arterial remodeling and right ventricular hypertrophy generated by hypoxia. In the research conducted by Akagi et al., they have shown that the administration of imatinib-incorporated nanoparticles into the trachea in rats may effectively impede the progression of pulmonary arterial hypertension (PAH) produced by monocrotaline (MCT) [46, 89, 126].

Lipid nanoparticles (LNPs) have attracted significant attention in the field of cardiac tissue engineering because of their capacity to serve as efficient carriers for siRNA delivery. The intriguing potential of using siRNA-encapsulated LNPs for the treatment of atherosclerosis and chronic inflammation in cardiovascular disease (CVD) has been shown by the delivery of these LNPs systemically. This strategy has shown persistent gene silencing effects for a duration of 10 days. Leuschner et al. demonstrated in their study that upon systemic treatment, lipidoid C12-200-LNPs encapsulating siRNA molecules designed to target the chemokine receptor CCR2 exhibited successful localization in the spleen and bone marrow. The process of localization resulted in the inhibition of the in vivo production of atherosclerotic plaque [71, 72, 76, 154].

2 Conclusion

In this chapter, we have explored the diverse spectrum of nanomaterials, particularly nanoparticles, and their wide-ranging applications within the field of tissue engineering. Recent strides in nanotechnology have yielded significant advantages for tissue engineering, a discipline focused on the repair and reconstruction of damaged tissues and organs, as well as the design of intelligent drug delivery systems. The development and utilization of nanomaterials in tissue engineering play a pivotal role in the restoration and regeneration of compromised tissues. In the context of contemporary nanotechnology, an increasing number of researchers are actively engaged in crafting novel biomaterials through various combinations of diverse nanomaterials. However, the integration of these nanomaterials into tissue engineering for organ replacement necessitates a judicious examination of factors such as the materials' sensitivity, potential immune responses, toxicity, impact on reproduction, and even fetal development. Such considerations are of paramount importance. The ongoing evolution of new nanomaterials presents an exciting opportunity for tissue engineering that must harmonize with the expectations of patients and the requirements of healthcare practitioners. Achieving significant strides in the biosafety, practicality, and durability of nanomaterials remains a significant challenge. We are optimistic that the rational design of future nanotechnologies will address many of the current challenges encountered in tissue engineering. In most scenarios, nanomaterial-based tissue regeneration has exhibited superior efficacy compared to traditional, artificial or animal-derived grafts. These conventional approaches often grapple with issues such as high costs, susceptibility to infection, inflammatory responses, immune reactions, and the need for periodic replacement. In general, nanomaterial-based tissue regeneration has delivered promising outcomes in the restoration and repair of damaged tissues, encompassing applications in areas like bone regeneration, skin rejuvenation, dental treatments, cardiovascular disorders, and neurological diseases.

References

1. Adams JC (2001) Cell-matrix contact structures. Cell Mol Life Sci. https://doi.org/10.1007/PL00000864
2. Ain QU et al (2017) Enhanced mechanical properties and biocompatibility of novel hydroxyapatite/TOPAS hybrid composite for bone tissue engineering applications. Mater Sci Eng, C. https://doi.org/10.1016/j.msec.2017.02.117
3. Al-Amri S et al (2014) Ni doped CuO nanoparticles: structural and optical characterizations. Curr Nanosci. https://doi.org/10.2174/1573413710666141024212856
4. Alghamdi H, Jansen J (2019) Dental implants and bone grafts: materials and biological issues, dental implants and bone grafts: materials and biological issues. https://doi.org/10.1016/C2016-0-03454-X
5. Backlund CJ et al (2015) Kinetic-dependent killing of oral pathogens with nitric oxide. J Dent Res. https://doi.org/10.1177/0022034515589314
6. Balasundaram G, Webster TJ (2006) Nanotechnology and biomaterials for orthopedic medical applications. Nanomedicine. https://doi.org/10.2217/17435889.1.2.169
7. Bao X et al (2018) Polydopamine nanoparticles as efficient scavengers for reactive oxygen species in periodontal disease. ACS Nano. https://doi.org/10.1021/acsnano.8b04022
8. Barry BW (2001) Novel mechanisms and devices to enable successful transdermal drug delivery. Eur J Pharm Sci. https://doi.org/10.1016/S0928-0987(01)00167-1
9. Bellu E et al (2021) Nanomaterials in skin regeneration and rejuvenation. Int J Mol Sci. https://doi.org/10.3390/ijms22137095
10. Benzaid R et al (2008) Fracture toughness, strength and slow crack growth in a ceria stabilized zirconia-alumina nanocomposite for medical applications. Biomaterials. https://doi.org/10.1016/j.biomaterials.2008.05.021
11. Besinis A, Van Noort R, Martin N (2012) Infiltration of demineralized dentin with silica and hydroxyapatite nanoparticles. Dent Mater. https://doi.org/10.1016/j.dental.2012.05.007

12. Chen H et al (2018) Magnetic cell-scaffold interface constructed by superparamagnetic IONP enhanced osteogenesis of adipose-derived stem cells. ACS Appl Mater Interfaces. https://doi.org/10.1021/acsami.8b17427
13. Coales P (2000) Principles of anatomy and physiology. Physiotherapy. https://doi.org/10.1016/s0031-9406(05)60992-3
14. Covarrubias C et al (2018) Bionanocomposite scaffolds based on chitosan–gelatin and nanodimensional bioactive glass particles: In vitro properties and in vivo bone regeneration. J Biomater Appl. https://doi.org/10.1177/0885328218759042
15. Danhier F et al (2012) PLGA-based nanoparticles: an overview of biomedical applications. J Control Release. https://doi.org/10.1016/j.jconrel.2012.01.043
16. Das S et al (2015) In vivo studies of silk based gold nano-composite conduits for functional peripheral nerve regeneration. Biomaterials. https://doi.org/10.1016/j.biomaterials.2015.04.047
17. Davis ME et al (2006) Local myocardial insulin-like growth factor 1 (IGF-1) delivery with biotinylated peptide nanofibers improves cell therapy for myocardial infarction. Proc Natl Acad Sci USA. https://doi.org/10.1073/pnas.0602877103
18. Denry I, Kelly JR (2008) State of the art of zirconia for dental applications. Dent Mater. https://doi.org/10.1016/j.dental.2007.05.007
19. Deville S et al (2003) Low-temperature ageing of zirconia-toughened alumina ceramics and its implication in biomedical implants. J Eur Ceram Soc. https://doi.org/10.1016/S0955-2219(03)00313-3
20. Deville S, Saiz E, Tomsia AP (2006) Freeze casting of hydroxyapatite scaffolds for bone tissue engineering. Biomaterials. https://doi.org/10.1016/j.biomaterials.2006.06.028
21. Dikpati, A. et al. (2012) 'Targeted Drug Delivery to CNS using Nanoparticles', Journal of Advanced Pharmaceutical Sciences.
22. Dong X (2018) Current strategies for brain drug delivery. Theranostics. https://doi.org/10.7150/thno.21254
23. Dvir T et al (2011) Nanowired three-dimensional cardiac patches. Nat Nanotechnol. https://doi.org/10.1038/nnano.2011.160
24. Ekerdt BL et al (2018) Thermoreversible hyaluronic acid-PNIPAAm hydrogel systems for 3D stem cell culture. Adv Healthcare Mater. https://doi.org/10.1002/adhm.201800225
25. Elsabahy M et al (2015) Polymeric nanostructures for imaging and therapy. Chem Rev. https://doi.org/10.1021/acs.chemrev.5b00135
26. Essaghraoui A et al (2019) Improved dermal delivery of cyclosporine a loaded in solid lipid nanoparticles. Nanomaterials. https://doi.org/10.3390/nano9091204
27. Estelrich J et al (2015) Iron oxide nanoparticles for magnetically-guided and magnetically-responsive drug delivery. Int J Mol Sci. https://doi.org/10.3390/ijms16048070
28. Fang Z et al (2020) Enhancement of sciatic nerve regeneration with dual delivery of vascular endothelial growth factor and nerve growth factor genes. J Nanobiotechnol. https://doi.org/10.1186/s12951-020-00606-5
29. Faustino C, Rijo P, Reis CP (2017) Nanotechnological strategies for nerve growth factor delivery: therapeutic implications in Alzheimer's disease. Pharmacol Res. https://doi.org/10.1016/j.phrs.2017.03.020
30. Feiner R et al (2016) Engineered hybrid cardiac patches with multifunctional electronics for online monitoring and regulation of tissue function. Nat Mater. https://doi.org/10.1038/nmat4590
31. Ferguson EL et al (2018) Controlled release of dextrin-conjugated growth factors to support growth and differentiation of neural stem cells. Stem Cell Res. https://doi.org/10.1016/j.scr.2018.10.008
32. Fu YC et al (2014) A novel single pulsed electromagnetic field stimulates osteogenesis of bone marrow mesenchymal stem cells and bone repair. PLoS ONE. https://doi.org/10.1371/journal.pone.0091581
33. Fujioka K et al (2014) Effects of silica and titanium oxide particles on a human neural stem cell line: morphology, mitochondrial activity, and gene expression of differentiation markers. Int J Mol Sci. https://doi.org/10.3390/ijms150711742

34. Gaharwar AK, Peppas NA, Khademhosseini A (2014) Nanocomposite hydrogels for biomedical applications. Biotechnol Bioeng. https://doi.org/10.1002/bit.25160

35 Geiser M et al (2005) Ultrafine particles cross cellular membranes by nonphagocytic mechanisms in lungs and in cultured cells. Environ Health Perspect. https://doi.org/10.1289/ehp.8006

36. Ghafar H et al (2020) Development and characterization of bioadhesive film embedded with lignocaine and calcium fluoride nanoparticles. AAPS PharmSciTech. https://doi.org/10.1208/s12249-019-1615-5

37. Glat M et al (2013) Age-dependent effects of microglial inhibition in vivo on Alzheimer's disease neuropathology using bioactive-conjugated iron oxide nanoparticles. J Nanobiotechnol. https://doi.org/10.1186/1477-3155-11-32

38. Gobin AM et al (2005) Near infrared laser-tissue welding using nanoshells as an exogenous absorber. Lasers Surg Med. https://doi.org/10.1002/lsm.20206

39. Godinho BMDC et al (2014) Differential nanotoxicological and neuroinflammatory liabilities of non-viral vectors for RNA interference in the central nervous system. Biomaterials. https://doi.org/10.1016/j.biomaterials.2013.09.068

40. Gonzalez-Carter D et al (2020) Targeting nanoparticles to the brain by exploiting the blood brain barrier impermeability to selectively label the brain endothelium. Proc Natl Acad Sci USA. https://doi.org/10.1073/pnas.2002016117

41. Größner-Schreiber B et al (2001) Plaque formation on surface modified dental implants—an in vitro study. Clin Oral Implant Res. https://doi.org/10.1034/j.1600-0501.2001.120601.x

42. Guo J et al (2017) Complexes of magnetic nanoparticles with cellulose nanocrystals as regenerable, highly efficient, and selective platform for protein separation. Biomacromol. https://doi.org/10.1021/acs.biomac.6b01778

43. Gupta A, Singh S (2022) Multimodal potentials of gold nanoparticles for bone tissue engineering and regenerative medicine: avenues and prospects. Small. https://doi.org/10.1002/smll.202201462

44. Hao T et al (2017) Injectable fullerenol/alginate hydrogel for suppression of oxidative stress damage in brown adipose-derived stem cells and cardiac repair. ACS Nano. https://doi.org/10.1021/acsnano.7b00221

45. Huey R et al (2017) Targeted drug delivery system to neural cells utilizes the nicotinic acetylcholine receptor. Int J Pharm. https://doi.org/10.1016/j.ijpharm.2017.04.023

46. Ishihara T et al (2015) Encapsulation of beraprost sodium in nanoparticles: analysis of sustained release properties, targeting abilities and pharmacological activities in animal models of pulmonary arterial hypertension. J Control Release: Off J Control Release Soc. https://doi.org/10.1016/j.jconrel.2014.10.029

47. Jin LJ et al (2016) Global burden of oral diseases: emerging concepts, management and interplay with systemic health. Oral Dis. https://doi.org/10.1111/odi.12428

48. Kalashnikova I, Das S, Seal S (2015) Nanomaterials for wound healing: Scope and advancement. Nanomedicine. https://doi.org/10.2217/nnm.15.82

49 Kandori K et al (2011) Preparation of calcium hydroxyapatite nanoparticles using microreactor and their characteristics of protein adsorption. J Phys Chem B. https://doi.org/10.1021/jp110441e

50. Khan HA et al (2019) Size and time-dependent induction of proinflammatory cytokines expression in brains of mice treated with gold nanoparticles. Saudi J Biol Sci. https://doi.org/10.1016/j.sjbs.2018.09.012

51. Khatami M et al (2018) Applications of green synthesized Ag, ZnO and Ag/ZnO nanoparticles for making clinical antimicrobial wound-healing bandages. Sustain Chem Pharm. https://doi.org/10.1016/j.scp.2018.08.001

52. Khetawat S (2015) Nanotechnology (Nanohydroxyapatite Crystals): recent advancement in treatment of dentinal hypersensitivity. JBR J Interdiscip Med Dent Sci. https://doi.org/10.4172/2376-032x.1000181

53. Khurshid Z et al (2015) Advances in nanotechnology for restorative dentistry. Materials. https://doi.org/10.3390/ma8020717

54. Kim JA et al (2011) Enhancement of neurite outgrowth in PC12 cells by iron oxide nanoparticles. Biomaterials. https://doi.org/10.1016/j.biomaterials.2011.01.019

55. Kim JJ et al (2014) Magnetic scaffolds of polycaprolactone with functionalized magnetite nanoparticles: physicochemical, mechanical, and biological properties effective for bone regeneration. RSC Adv. https://doi.org/10.1039/c4ra00040d

56. Kim JS et al (2007) Antimicrobial effects of silver nanoparticles. Nanomed Nanotechnol Biol Med. https://doi.org/10.1016/j.nano.2006.12.001

57. Kim K, Fisher JP (2007) Nanoparticle technology in bone tissue engineering. J Drug Target. https://doi.org/10.1080/10611860701289818

58. Kim T et al (2013) Graphene films show stable cell attachment and biocompatibility with electrogenic primary cardiac cells. Mol Cells. https://doi.org/10.1007/s10059-013-0277-5

59. Ko WK et al (2015) The effect of gold nanoparticle size on osteogenic differentiation of adipose-derived stem cells. J Colloid Interface Sci. https://doi.org/10.1016/j.jcis.2014.08.058

60. Kong SD et al (2012) Magnetic targeting of nanoparticles across the intact blood-brain barrier. J Control Release. https://doi.org/10.1016/j.jconrel.2012.09.021

61. Korrapati PS et al (2016) Recent advancements in nanotechnological strategies in selection, design and delivery of biomolecules for skin regeneration. Mater Sci Eng, C. https://doi.org/10.1016/j.msec.2016.05.074

62. Kotani H et al (2002) Strong static magnetic field stimulates bone formation to a definite orientation in vitro and in vivo. J Bone Miner Res. https://doi.org/10.1359/jbmr.2002.17.10.1814

63. Kudr J et al (2017) Magnetic nanoparticles: from design and synthesis to real world applications. Nanomaterials. https://doi.org/10.3390/nano7090243

64. Kulshrestha S et al (2014) A graphene/zinc oxide nanocomposite film protects dental implant surfaces against cariogenic Streptococcus mutans. Biofouling. https://doi.org/10.1080/08927014.2014.983093

65. Kumar N et al (2022) Advanced metal and carbon nanostructures for medical, drug delivery and bio-imaging applications. Nanoscale. https://doi.org/10.1039/d1nr07643d

66. De La Escosura-Muñiz A, Merkoçi A (2012) Nanochannels preparation and application in biosensing. ACS Nano. https://doi.org/10.1021/nn301368z

67. Landsiedel R et al (2010) Gene toxicity studies on titanium dioxide and zinc oxide nanomaterials used for UV-protection in cosmetic formulations. Nanotoxicology. https://doi.org/10.3109/17435390.2010.506694

68. Laurent S et al (2008) Magnetic iron oxide nanoparticles: synthesis, stabilization, vectorization, physicochemical characterizations and biological applications. Chem Rev. https://doi.org/10.1021/cr068445e

69. Lee HR et al (2013) Comparative characteristics of porous bioceramics for an osteogenic response in vitro and in vivo. PLoS ONE. https://doi.org/10.1371/journal.pone.0084272

70. Lee JKY et al (2018) Polymer-based composites by electrospinning: preparation & functionalization with nanocarbons. Prog Polym Sci. https://doi.org/10.1016/j.progpolymsci.2018.07.002

71. Leung AKK, Tam YYC, Cullis PR (2014) Lipid nanoparticles for short interfering RNA delivery. Adv Genet. https://doi.org/10.1016/B978-0-12-800148-6.00004-3

72. Leuschner F et al (2011) Therapeutic siRNA silencing in inflammatory monocytes in mice. Nat Biotechnol. https://doi.org/10.1038/nbt.1989

73. Lin C et al (2011) Incorporation of carboxylation multiwalled carbon nanotubes into biodegradable poly(lactic-co-glycolic acid) for bone tissue engineering. Colloids Surf B: Biointerfaces. https://doi.org/10.1016/j.colsurfb.2010.12.011

74. Liu J et al (2004) Peripherally delivered glutamic acid decarboxylase gene therapy for spinal cord injury pain. Mol Ther. https://doi.org/10.1016/j.ymthe.2004.04.017

75. Liu Q, Wang X, Yi S (2018) Pathophysiological changes of physical barriers of peripheral nerves after injury. Front Neurosci. https://doi.org/10.3389/fnins.2018.00597

76. Love KT et al (2010) Lipid-like materials for low-dose, in vivo gene silencing. Proc Natl Acad Sci USA. https://doi.org/10.1073/pnas.0910603106

77. Lu C, Fuchs E (2014) Sweat gland progenitors in development, homeostasis, and wound repair. Cold Spring Harb Perspect Med. https://doi.org/10.1101/cshperspect.a015222
78. Ma X et al (2019) Precisely controllable hybrid graphene scaffold reveals size effects on differentiation of neural progenitor cells in mimicking neural network. Carbon. https://doi.org/10.1016/j.carbon.2019.01.006
79. Manatunga DC et al (2020) Nanofibrous cosmetic face mask for transdermal delivery of nano gold: synthesis, characterization, release and zebra fish employed toxicity studies: nanofibrous mask for delivery of gold. R Soc Open Sci. https://doi.org/10.1098/rsos.201266
80. Martinelli V et al (2012) Carbon nanotubes promote growth and spontaneous electrical activity in cultured cardiac myocytes. Nano Lett. https://doi.org/10.1021/nl204064s
81. Masoudipour E et al (2018) A novel intracellular pH-responsive formulation for FTY720 based on PEGylated graphene oxide nano-sheets. Drug Dev Ind Pharm. https://doi.org/10.1080/03639045.2017.1386194
82. McBain SC, Yiu HHP, Dobson J (2008) Magnetic nanoparticles for gene and drug delivery. Int J Nanomed. https://doi.org/10.2147/ijn.s1608
83. McIntyre RA (2012) Common nano-materials and their use in real world applications. Sci Prog. https://doi.org/10.3184/003685012X13294715456431
84. Meng X et al (2013) Novel injectable biomimetic hydrogels with carbon nanofibers and self assembled rosette nanotubes for myocardial applications. J Biomed Mater Res-Part A. https://doi.org/10.1002/jbm.a.34400
85. Misch CE (2014) Dental implant prosthetics. Dent Implant Prosthet. https://doi.org/10.1016/C2009-0-42440-1
86. Moniz T, Costa Lima SA, Reis S (2020) Application of the human stratum corneum lipid-based mimetic model in assessment of drug-loaded nanoparticles for skin administration. Int J Pharm. https://doi.org/10.1016/j.ijpharm.2020.119960
87. Moraes G, Zambom C, Siqueira WL (2021) Nanoparticles in dentistry: a comprehensive review. Pharmaceuticals. https://doi.org/10.3390/ph14080752
88. Motte L (2012) What are the current advances regarding iron oxide nanoparticles for nanomedicine? J Bioanal Biomed. https://doi.org/10.4172/1948-593X.1000e110
89. Nakamura K et al (2021) Inhibitory effects of RAGE-aptamer on development of monocrotaline-induced pulmonary arterial hypertension in rats. J Cardiol. https://doi.org/10.1016/j.jjcc.2020.12.009
90. Neel EAA et al (2015) Nanotechnology in dentistry: prevention, diagnosis, and therapy. Int J Nanomed. https://doi.org/10.2147/IJN.S86033
91. Nguyen S et al (2017) Fluoride loaded polymeric nanoparticles for dental delivery. Eur J Pharm Sci. https://doi.org/10.1016/j.ejps.2017.04.004
92. Niza E et al (2019) Assessment of doxorubicin delivery devices based on tailored bare polycaprolactone against glioblastoma. Int J Pharm. https://doi.org/10.1016/j.ijpharm.2018.12.079
93. Oliveira JM et al (2006) Novel hydroxyapatite/chitosan bilayered scaffold for osteochondral tissue-engineering applications: scaffold design and its performance when seeded with goat bone marrow stromal cells. Biomaterials. https://doi.org/10.1016/j.biomaterials.2006.07.034
94. Pacholski C et al (2005) Biosensing using porous silicon double-layer interferometers: Reflective interferometric Fourier transform spectroscopy. J Am Chem Soc. https://doi.org/10.1021/ja0511671
95. Pakulska MM, Tator CH, Shoichet MS (2017) Local delivery of chondroitinase ABC with or without stromal cell-derived factor 1α promotes functional repair in the injured rat spinal cord. Biomaterials. https://doi.org/10.1016/j.biomaterials.2017.04.016
96. Parani M et al (2016) Engineered nanomaterials for infection control and healing acute and chronic wounds. ACS Appl Mater Interfaces. https://doi.org/10.1021/acsami.6b00291
97. Pattni BS, Chupin VV, Torchilin VP (2015) New developments in liposomal drug delivery. Chem Rev. https://doi.org/10.1021/acs.chemrev.5b00046
98. Paulose M et al (2006) Anodic growth of highly ordered TiO_2 nanotube arrays to 134 μm in length. J Phys Chem B. https://doi.org/10.1021/jp064020k

99. Pecharromán C et al (2003) Percolative mechanism of aging in zirconia-containing ceramics for medical applications. Adv Mater. https://doi.org/10.1002/adma.200390117

100. Perumal G et al (2017) Synthesis and characterization of curcumin loaded PLA—hyperbranched polyglycerol electrospun blend for wound dressing applications. Mater Sci Eng C. https://doi.org/10.1016/j.msec.2017.03.200

101. Pilakka-Kanthikeel S et al (2013) Targeted brain derived neurotropic factors (BDNF) delivery across the blood-brain barrier for neuro-protection using magnetic nano carriers: an in-vitro study. PLoS ONE. https://doi.org/10.1371/journal.pone.0062241

102. Poduslo JF et al (2011) Targeting vascular amyloid in arterioles of alzheimer disease transgenic mice with amyloid β protein antibody-coated nanoparticles. J Neuropathol Exp Neurol. https://doi.org/10.1097/NEN.0b013e318225038c

103. Priyadarsini S et al (2019) Application of nanoparticles in dentistry: current trends. Nanoparticles Med. https://doi.org/10.1007/978-981-13-8954-2_3

104. Qazi TH et al (2014) Development and characterization of novel electrically conductive PANI-PGS composites for cardiac tissue engineering applications. Acta Biomaterialia 10(6):2434–2445. . Acta Materialia Inc. https://doi.org/10.1016/j.actbio.2014.02.023

105. Amin RD et al (2020) Nanomaterials for cardiac tissue engineering. Molecules (Basel, Switzerland). https://doi.org/10.3390/molecules25215189

106. Rad AA et al (2019) The application of nanomaterials in cardiovascular diseases: a review on drugs and devices. J Pharm Pharm Sci. https://doi.org/10.18433/jpps30456

107. Rhee JW, Wu JC (2013) Advances in nanotechnology for the management of coronary artery disease. Trends Cardiovasc Med. https://doi.org/10.1016/j.tcm.2012.08.009

108. Van Rijt S, Habibovic P (2017) Enhancing regenerative approaches with nanoparticles. J R Soc Interface. https://doi.org/10.1098/rsif.2017.0093

109. Rink JS et al (2013) Update on current and potential nanoparticle cancer therapies. Curr Opin Oncol. https://doi.org/10.1097/CCO.0000000000000012

110. Saraiva C et al (2016) Nanoparticle-mediated brain drug delivery: overcoming blood-brain barrier to treat neurodegenerative diseases. J Control Release. https://doi.org/10.1016/j.jconrel.2016.05.044

111. Sarker MD et al (2018) Regeneration of peripheral nerves by nerve guidance conduits: influence of design, biopolymers, cells, growth factors, and physical stimuli. Prog Neurobiol. https://doi.org/10.1016/j.pneurobio.2018.07.002

112. Sathyanarayanan MB et al (2013) The effect of gold and iron-oxide nanoparticles on biofilm-forming pathogens. ISRN Microbiol. https://doi.org/10.1155/2013/272086

113. Sen CK et al (2009) Human skin wounds: a major and snowballing threat to public health and the economy: perspective article. Wound Repair Regen. https://doi.org/10.1111/j.1524-475X.2009.00543.x

114. Senthil Kumar R et al (2017) Nanochitosan modified glass ionomer cement with enhanced mechanical properties and fluoride release. Int J Biol Macromol. https://doi.org/10.1016/j.ijbiomac.2017.05.120

115. Shalaby MA, Anwar MM, Saeed H (2022) Nanomaterials for application in wound Healing: current state-of-the-art and future perspectives. J Polym Res. https://doi.org/10.1007/s10965-021-02870-x

116. Shankarappa SA, Sagie I et al (2012) Duration and local toxicity of sciatic nerve blockade with coinjected site 1 sodium-channel blockers and quaternary lidocaine derivatives. Reg Anesth Pain Med. https://doi.org/10.1097/AAP.0b013e31826125b3

117. Shankarappa SA, Tsui JH et al (2012) Prolonged nerve blockade delays the onset of neuropathic pain. Proc Natl Acad Sci USA. https://doi.org/10.1073/pnas.1214634109

118. Shao F et al (2018) Bio-synthesis of Barleria gibsoni leaf extract mediated zinc oxide nanoparticles and their formulation gel for wound therapy in nursing care of infants and children. J Photochem Photobiol B: Biol. https://doi.org/10.1016/j.jphotobiol.2018.10.014

119. Shao XR et al (2015) Independent effect of polymeric nanoparticle zeta potential/surface charge, on their cytotoxicity and affinity to cells. Cell Prolif. https://doi.org/10.1111/cpr.12192

120. Shuai C et al (2020) A magnetic micro-environment in scaffolds for stimulating bone regeneration. Mater Des. https://doi.org/10.1016/j.matdes.2019.108275
121 Singh RK et al (2014) Potential of magnetic nanofiber scaffolds with mechanical and biological properties applicable for bone regeneration. PLoS ONE. https://doi.org/10.1371/journal.pone.0091584
122. Son KH, Hong JH, Lee JW (2016) Carbon nanotubes as cancer therapeutic carriers and mediators. Int J Nanomed. https://doi.org/10.2147/IJN.S112660
123. Stoimenov PK et al (2002) Metal oxide nanoparticles as bactericidal agents. Langmuir. https://doi.org/10.1021/la0202374
124. Surnar B et al (2018) Nanotechnology-mediated crossing of two impermeable membranes to modulate the stars of the neurovascular unit for neuroprotection. Proc Natl Acad Sci USA. https://doi.org/10.1073/pnas.1816429115
125. Tajdaran K et al (2016) An engineered biocompatible drug delivery system enhances nerve regeneration after delayed repair. J Biomed Mater Res-Part A. https://doi.org/10.1002/jbm.a.35572
126. Takeda M et al (2009) Synthesis of prostaglandin E1 phosphate derivatives and their encapsulation in biodegradable nanoparticles. Pharm Res. https://doi.org/10.1007/s11095-009-9891-5
127. Talal A et al (2013) Effects of hydroxyapatite and PDGF concentrations on osteoblast growth in a nanohydroxyapatite-polylactic acid composite for guided tissue regeneration. J Mater Sci—Mater Med. https://doi.org/10.1007/s10856-013-4963-9
128. Tang JN et al (2018) Concise review: is cardiac cell therapy dead? Embarrassing trial outcomes and new directions for the future. Stem Cells Transl Med. https://doi.org/10.1002/sctm.17-0196
129. Versiani AF et al (2016) Gold nanoparticles and their applications in biomedicine. Futur Virol. https://doi.org/10.2217/fvl-2015-0010
130. Vieira S et al (2017) Nanoparticles for bone tissue engineering. Biotechnol Prog. https://doi.org/10.1002/btpr.2469
131. Wahajuddin and Arora S (2012) Superparamagnetic iron oxide nanoparticles: magnetic nanoplatforms as drug carriers. Int J Nanomed. https://doi.org/10.2147/IJN.S30320
132. Wang DK, Rahimi M, Filgueira CS (2021) Nanotechnology applications for cardiovascular disease treatment: current and future perspectives. Nanomed Nanotechnol Biol Med. https://doi.org/10.1016/j.nano.2021.102387
133. Wang P et al (2015) Bone tissue engineering via nanostructured calcium phosphate biomaterials and stem cells. Bone Res. https://doi.org/10.1038/boneres.2014.17
134. Webster TJ, Tran (2011) Selenium nanoparticles inhibit Staphylococcus aureus growth. Int J Nanomed. https://doi.org/10.2147/ijn.s21729
135. Wei M, Li S, Le W (2017) Nanomaterials modulate stem cell differentiation: biological interaction and underlying mechanisms. J Nanobiotechnol. https://doi.org/10.1186/s12951-017-0310-5
136. Wolfe D, Mata M, Fink DJ (2012) Targeted drug delivery to the peripheral nervous system using gene therapy. Neurosci Lett. https://doi.org/10.1016/j.neulet.2012.04.047
137. Wolfram JA, Donahue JK (2013) Gene therapy to treat cardiovascular disease. J Am Hear Assoc. https://doi.org/10.1161/JAHA.113.000119
138. Wood NJ et al (2015) Chlorhexidine hexametaphosphate nanoparticles as a novel antimicrobial coating for dental implants. J Mater Sci—Mater Med. https://doi.org/10.1007/s10856-015-5532-1
139. Xia Y et al (2019) Iron oxide nanoparticle-calcium phosphate cement enhanced the osteogenic activities of stem cells through WNT/β-catenin signaling. Mater Sci Eng C. https://doi.org/10.1016/j.msec.2019.109955
140. Xing ZC et al (2010) In vitro assessment of antibacterial activity and cytocompatibility of silver-containing phbv nanofibrous scaffolds for tissue engineering. Biomacromol. https://doi.org/10.1021/bm1000372

141. Xiu ZM, Ma J, Alvarez PJJ (2011) Differential effect of common ligands and molecular oxygen on antimicrobial activity of silver nanoparticles versus silver ions. Environ Sci Technol. https://doi.org/10.1021/es201918f
142. Yang H et al (2019) An in Vivo miRNA delivery system for restoring infarcted myocardium. ACS Nano. https://doi.org/10.1021/acsnano.9b03343
143. Yang X et al (2015) Gold nanomaterials at work in biomedicine. Chem Rev. https://doi.org/10.1021/acs.chemrev.5b00193
144. Yang X et al (2017) Pharmaceutical intermediate-modified gold nanoparticles: against multidrug-resistant bacteria and wound-healing application via an electrospun scaffold. ACS Nano. https://doi.org/10.1021/acsnano.7b01240
145. Yao Y et al (2018) Potentials of combining nanomaterials and stem cell therapy in myocardial repair. Nanomedicine. https://doi.org/10.2217/nnm-2018-0013
146. Zabielska-Koczywąs K, Lechowski R (2017) The use of liposomes and nanoparticles as drug delivery systems to improve cancer treatment in dogs and cats. Molecules. https://doi.org/10.3390/molecules22122167
147. Zafar MS, Najeeb S, Khurshid Z (2020) 'Introduction to dental implants materials, coatings, and surface modifications. Dent Implant: Mater Coat Surf Modif Interfaces Oral Tissues. https://doi.org/10.1016/B978-0-12-819586-4.00001-9
148. Zamproni LN et al (2017) Injection of SDF-1 loaded nanoparticles following traumatic brain injury stimulates neural stem cell recruitment. Int J Pharm. https://doi.org/10.1016/j.ijpharm.2017.01.036
149. Zhang D et al (2014) Gold nanoparticles stimulate differentiation and mineralization of primary osteoblasts through the ERK/MAPK signaling pathway. Mater Sci Eng C. https://doi.org/10.1016/j.msec.2014.04.042
150. Zhang R et al (2016) Traceable nanoparticle delivery of small interfering rna and retinoic acid with temporally release ability to control neural stem cell differentiation for alzheimer's disease therapy. Adv Mater. https://doi.org/10.1002/adma.201600554
151. Zheng X et al (2021) Applications of nanomaterials in tissue engineering. RSC Adv. https://doi.org/10.1039/d1ra01849c
152. Zhou W et al (2015) Gold nanoparticles for in vitro diagnostics. Chem Rev. https://doi.org/10.1021/acs.chemrev.5b00100
153. Zhou W et al (2016) Tuning the mechanical properties of Poly(Ethylene Glycol) microgel-based scaffolds to increase 3D schwann cell proliferation. Macromol Biosci. https://doi.org/10.1002/mabi.201500336
154. Zimmermann TS et al (2006) RNAi-mediated gene silencing in non-human primates. Nature. https://doi.org/10.1038/nature04688
155. Zuidema JM et al (2015) Magnetic NGF-releasing PLLA/Iron oxide nanoparticles direct extending neurites and preferentially guide neurites along aligned electrospun microfibers. ACS Chem Neurosci. https://doi.org/10.1021/acschemneuro.5b00189

Nanostructured Titanium Surfaces in Hard Tissue Repair

Eylül Yakar, Boğaç Kılıçarslan, and Cem Bayram

Abstract Since hard tissues such as bone have a hierarchical structure with properties on many length scales from nano-level to macro-level, nanostructured titanium surfaces play an important role in hard tissue repair because they can mimic natural bone. Nanostructured surfaces can mimic the nanoscale properties of natural bone that promote bone cell adhesion, proliferation, and differentiation, as well as increase the surface area available for cell attachment, thus the contact area between the implant and the surrounding bone, improving the stability of the implant and its integration with the host tissue. Nanostructured titanium surfaces can also facilitate deposition of mineralized matrix and promote bone regeneration. In addition, the use of nanostructured surfaces can reduce the risk of implant failure and the requirement for additional surgical procedures. They can improve the long-term clinical outcomes of bone implants and reduce the likelihood of complications such as infection, inflammation, and rejection. Nanostructured titanium surfaces have significant potential to increase the efficiency of hard tissue repair and improve the quality of life of patients with bone injuries or disorders.

Keywords Titanium · Titania nanotubes · Osseointegration · Anodic oxidation · Implant · Surface modification · Antibacterial surface · Drug eluting implant

EY and BK equally contributed to this work.

E. Yakar · B. Kılıçarslan · C. Bayram (✉)
Department of Nanotechnology and Nanomedicine, Graduate School of Science and Technology, Hacettepe University, Ankara, Turkey
e-mail: cemb@hacettepe.edu.tr

27

1 Introduction

The aging population, accidents, and injuries worldwide have contributed to an increase in demand for hard tissue implants and fixation devices. These devices refer to materials used in cases where there is significant reduction and dysfunction in bone mass, and they are expected to restore the patient's healthy vital movements and activities promptly and efficiently. Titanium, CoCrMo (cobalt, chromium, molybdenum), and 316L stainless steel are commonly utilized in this material category. The mechanical features of these materials resemble most to natural tissue, and their average implant life is expected to last between 15 and 20 years. Although this lifespan is relatively acceptable for the elderly population, it falls short of the needs of younger individuals who require implants and fixation. In addition to replacing lost bone tissue, a bone implant ought to serve as a platform for the regeneration and healing of bone and vascular tissue in the surrounding local tissue. It is primordial that the implant's point of contact with the affected tissue encourages cell adhesion, proliferation, and differentiation, in simple terms, the creation of new bone. By advancing new bone formation and vascularization, the implant and the bone are able to establish solid connections, commonly referred to as fracture-crack or loss.

While numerous factors impact the success of implantation and ossification, surgical technique, implant design, surface topography, surface chemistry, and wettability emerge as the primary features in the development of state-of-the-art implant devices. It is essential to recognize these aspects when striving to enhance the performance and longevity of the latest generation of implant devices. Although the bulk and mechanical properties are of paramount importance in a bone implant, the surface properties of the material are paramount in the interaction with the tissue, as direct cytocompatibility is, by definition, a cell-surface event. Cells are sensitive to the topography and chemistry on the implant material. Topography is a crucial factor in both cell proliferation and differentiation. Surface roughness, a subcategory of topography, has undergone significant research for enhancing implant yield, and it has been found that surfaces with micropores ($1–100$ μm), produced by sandblasting and abrasion, boost osteoblast function when compared to smooth surfaces. Enhanced functions include initial cell adhesion, proliferation, alkaline phosphatase activity, and calcium mineral deposition. Products with surfaces acquired through these techniques have established their place in the current implant market.

It is widely understood that micro-rough surfaces do not perfectly align with the natural structure of hard tissue. Consequently, despite the previously mentioned advancements, implant efficiency and osteoinduction have not reached optimal levels. Collagen fibrils and hydroxyapatite mineral, which are the fundamental natural components of bone, exhibit a nanoscale structure. The diameter of these fibrils is significant. The fibrillar collagen bundles have a diameter of around 300 nm and make up 90% of the protein content in the extracellular matrix. Additionally, the hydroxyapatite crystal accounts for the majority of the inorganic aspect and is measured at $4 \times 25 \times 50$ nm. Ensuring the implant material surface is structured at this level in terms of size will inevitably lead to better success rates in the bone-material overlap.

When examining the modifications made to implant surfaces composed of titanium and its alloys, it becomes evident that only morphological changes were initially carried out. Direct surface modifications of titanium at nano-level generally associated with the process of anodic oxidation, which constitutes solely a morphological alteration. For surface treatments to be designated as "biomimetic", they must simulate the constituent and/or formal environment of the natural environment. Deposition of inorganic calcium phosphate salts onto titanium surfaces, plasma-assisted coatings, and those generated via sol–gel technique are prominent biomimetic modifications. Since the introduction of anodic oxidation into material modifications of implants after the latter half of the 2000s, modifications of implant surfaces on the nano-level have gained momentum.

The principal concern in designing and synthesizing implant biomaterials is to replicate the characteristics of natural bone. Synthetic materials from the past have been unable to deliver entirely gratifying outcomes. The first generation of hard tissue implants, such as those in the hip, knee, teeth, and elbow, has a typical lifespan of 15–20 years. Conventional materials, or components of these materials in the micron size range, lack the potential to trigger the cellular response necessary for bone regeneration essential for long-term usage. In contrast, the implementation of nanophase materials, which can emulate bone structure components such as proteins and hydroxyapatite at the dimensional level, could provide an alternative approach for more effective hard tissue implants.

The National Nanotechnology Initiative in the United States provides a definition of nanotechnology as "the controlled manipulation of materials at the atomic, molecular and supramolecular level for imaging, measurement, research, and fabrication purposes". It is widely accepted that materials studied at this scale usually have dimensions between approximately 1 and 100 nm. Such materials exhibit physical, chemical, or biological responses that differ from their bulk counterparts. The physical and chemical properties of nanoscale materials in bulk structures can undergo changes which result in different responses like better electrical or optical performance. They can also gain higher structural integrity and more reactive character due to their increasing surface area.

Firstly, it is essential to clarify why there is a need for improved hard tissue implants, with the aim of emphasizing the significance of nanophase materials as an effective development pathway. In the subsequent sections, an examination will be made of the vital function of nanostructuring in manipulating cell functions and the effective employment of nanophase materials to address the obstacles faced in hard tissue regeneration. In addition to tissue development, bacterial infection is a common issue associated with implant use post-implantation. This chapter also summarizes the existence of antibacterial properties offered by nanostructured surfaces with their morpho-physical properties and the specialization of some specific nanostructured surfaces as structures that can release drugs locally.

2 Fabrication of Nanostructured Surfaces on Titanium

Titanium metal is shielded by a thin oxide layer that self-generates when exposed to oxygen in the environment. The oxide layer, which is roughly 2–5 nm in thickness, is accountable for the widely known resistance to corrosion of both titanium and titanium alloys [1]. As previously stated, titanium is a prevalent choice for hard tissue implants owing to its biocompatibility and mechanical properties. However, the naturally occurring oxide layer on the material's surface may not be active enough to facilitate an anchoring mechanism with hard tissue, potentially leading to implant failure [2–4].

The surface of titanium for use in implants has been the subject of much research and development, with examples including mechanical approaches (such as sandblasting), chemical treatments (including acid etching), and coatings (plasma spray). These techniques can improve the topography, chemistry and surface energy of the titanium [5–8]. As a result of these standard methods, implants that can more effectively bond with cells have been developed. This has been achieved by creating optimal micro-roughness, developing a feasible surface character, and achieving morphology that is preferred by bone cells. However, complete surface topography control cannot be guaranteed with any of the aforementioned methods. Additionally, it is possible that residues may remain on the surface due to the methods used. As a result, alternative surface modification techniques are required.

In conjunction with direct modifications, frequently employed techniques include coating the surface with hydroxyapatite or other calcium phosphate salts to enhance bonding with bone through simulated body fluid or plasma-spray methods [9–11]. However, coating efficiency may decrease over time if the inorganic materials do not adhere sufficiently to the surface.

2.1 Direct Manipulation of Titanium

The discovery of anodic oxidation of titanium dates back to 1930s, and since 1960s, it has been studied to enhance osseointegration in orthopedic implants. Studies employing the micro-arc oxidation technique have led to the formation of a microporous surface. This technique relies on a dielectric barrier formed on the anode surface by the new oxide layer, which grows and thickens until it reaches the dielectric breakdown point. The anodizing layer typically lacks uniformity because it contains cracks and defects caused by the uneven deposition of the oxide layer. As the applied potential increases, arc discharges take place, leading to dielectric collapse at weak points. This results in high temperatures at these points, which may reach several thousand Kelvin degrees, and localized surface melting may occur as well. As weak points increase on the surface, this thermal effect causes the above mechanism to repeat in smaller localized areas.

Titania layers acquired through micro-arc oxidation exhibit a rough and porous texture along with micrometer-level cracks [12]. The pore sizes tend to vary, depending on the process parameters, and the overall structure is not uniform. The pores range from a few hundred nanometers to a few microns in diameter, and all cavities and pores connect with each other. While the pores present are mostly round shaped, they are not uniform. Studies have shown that the pore diameter and film roughness generally increase in proportion to current density, applied potential, and electrolyte concentration [13, 14].

Anodization, commonly referred to as anodic oxidation, is an electrochemical technique that produces protective oxide layers on surfaces of metals. This process has facilitated the development of new surface coatings on implants consisting of valve metal, particularly over the past two decades. The technique allows physico-chemical and morphological modification of the surface character of titanium and its alloys inserted in physiological surrounding without interfering with their bulk structure.

A standard anodic oxidation process comprises alkaline cleaning, acid activation, and electrolytic anodization. Organic impurities on the surface must be removed during the cleaning stage prior to activation step. An activation phase follows, which involves an acid etch using a mixture of HNO_3 and HF to eliminate the naturally formed oxide layer. The process of anodization occurs within an electrochemical cell whereby titanium metal serves as the anode, and platinum functions as the cathode. An oxide layer is produced on the anode surface due to oxidation and reduction reactions that is the result of an applied constant potential voltage between the two electrodes [15, 16].

The characteristics of the oxide layer produced through anodization, including nanoscale roughness, morphology, and chemical composition, can fluctuate exten-sively depending on various parameters of the process including applied potential and current density or physicochemical properties like composition, pH, or tempera-ture. Both inorganic and organic acids, neutral salt, or alkaline solutions are common electrolytes utilized in anodic oxidation in the anodization of titanium. In a pioneer work, Sul et al. [17] conducted a comprehensive study on electrochemical oxide layer development. The researchers observed that processes using sulfuric acid solu-tion had the highest thickness of anodic oxidation among commonly used elec-trolytes. It is widely acknowledged that acidic electrolytes possess a greater ability to form oxide layer in comparison to alkaline electrolytes. When using electrolytes based on phosphoric acid and sulfuric acid at high voltages, layers that are tens of microns thick with a micrometer level of roughness are produced. In contrast, recent studies utilizing solutions containing fluoride ions have reported the achievement of nanotubular structures that are biologically compatible. The process of anodization can be executed through either a potentiostatic or galvanostatic reaction. During the procedure, when the applied potential exceeds the dielectric breakdown value of the oxide, current flow will not continue to develop the oxide layer, resulting in arcing discharges from the layer. It has been determined that the values for precipitation of phosphoric acid and sulfuric acid are 80 and 100 V, correspondingly [18]. When the precipitation limit is reached, the anodic oxide layer becomes thin and usually non-porous.

Temperature is another process parameter that affects oxide formation. For a uniform surface, a constant temperature is preferred throughout the process. While increasing temperature accelerates the dissolution of the oxide layer, resulting in increased porosity, it is advisable to keep it low so as to prevent the layer from completely separating from the surface.

The growth of the oxide film is the result of a balance between the creation of an oxide film and its dissolution, which is contingent on the electrolyte's impact. The impact of the electrolyte varies with additional system parameters like pH, concentration, applied potential, and current density. While the production of microporous titanium surfaces is a familiar process, previous studies on anodization have mainly concentrated on nanoscale levels, particularly to improve the biocompatibility of titanium biomaterials and achieve a surface that mimics subcellular dimensions. Nanoporous surfaces were produced on titanium through anodization between 10 and 40 V, utilizing a fluoride-containing electrolyte with voltages below the dielectric breakdown threshold. During the investigations, oxide layers in the shape of tubes measuring between 60 and 4400 nm in thickness and 20 and 500 nm in diameter were measured within these restricted parameters.

The first report on the necessity of fluoride ions in the creation of nanoporous titania structures at low voltages came from the study carried out by Zwilling et al. [19]. Subsequently, Gong et al. [20] reported the formation of nanotubular structures through anodization of titanium in a dilute hydrofluoric acid solution (0.5–1.5% by weight) at voltages ranging from 10 to 40 V. The study found that the diameters of the nanotube structures increased with the applied voltage, while the layer thickness, or nanotube length, increased with anodization time. Beranek et al. similarly obtained nanotube layers measuring 140 nm in diameter and 540 nm in thickness after 24 h of anodization using the HF/H_2SO_4 electrolyte system [21]. Comparable surfaces can also be achieved with organic electrolytes. Nanotubular structures were synthesized in Tsuchiya's study using an electrolyte composed of anhydrous mixtures of acetic acid and ammonium fluoride [22]. These studies showed that the oxide layer thickness did not exceed a few hundred nanometers. Other literature studies have reported titania oxide layers with thicknesses up to several micrometers, using various electrolytes. In the anodization studies carried out by Cai et al., it was observed that the thickness of the titanium oxide layer exceeded 4 μm when using KF and NaF aqueous solutions with a pH of 4.5 [23]. Similarly, the group reported a thickness of 2.3 μm for the titania surface when employing DMSO/ethanol/HF electrolyte [24].

Fluoride electrode anodization primarily involves field-assisted chemical dissolution, and subsequent oxidation. Among these, field-assisted dissolution is acknowledged as the principal mechanism in titania nanotube structure formation. Mor et al. elucidated the formation process of titania nanotubes using the point defect model. Initial pore formation occurs through localized dissolution at vulnerable areas, with non-anodized metal components also dispersed in the pores [25]. Small voids are subsequently formed in the regions between these pores through field-assisted oxidation/dissolution. Once the development of these voids attains equilibrium with the pores, the final shapes of the nanotube surface forms are adopted.

In contrast with micro-arc oxidation, nanotubular titanium oxide surfaces remain highly structured, even with increased applied voltage. The range of pore diameter is restricted to a few tens of nanometers to 100–200 nm and rises directly with the applied voltage [26]. The thickness of nanotube structures on the surface is a few hundred nanometers, which may be expanded to several micrometers by managing the pH and electrolyte composition. While the entirety of the structure is deemed to be uniform, any surface defect on the substrate surface may induce relative changes in the nanotube structures.

2.2 Nanostructured Coatings

The significance of nanotopographic surface modifications for osseointegration is widely recognized. Nonetheless, given the circumstances that titanium or its alloys subject to electrochemical surface adjustments face in a physiological milieu, which usually include organic tissues and bodily fluids, it is imperative to have surfaces that can organically and dimensionally imitate these living tissues. The adhesion, proliferation, and differentiation of cells in the microenvironment at the implant–tissue interface can be improved or directed using this approach. Nanotopography may provide dimensional improvements that enhance integration and expedite organic coating pathways. Inorganic coatings are also successful in surface modification of titanium. Firstly, the hydroxyapatite structure is the predominant component of the bone structure. To replicate the natural environment of bone cells, researchers have conducted successful studies utilizing nanoscale crystals and a rough calcium phosphate derivative structure on titanium metal. These findings have been documented in the literature. Until today, numerous studies have investigated the hypothesis that the osseointegrative ability of titanium and its alloys may be improved by coating techniques. Thus, early implant failure and long-term revision surgery problems can be prevented [27]. There are three important points as success criteria for implant coatings. The first of these, which is the most important element of the cause–effect relationship, is that "it must have a feature that enables osseointegration and bone development in the early period" and the second is that this coating must exhibit a very strong adhesion to the implant surface. Since it is necessary not to detach or peel off from the surfaces during harsh implantation conditions, it may be necessary to apply a pre-treatment with chemical surface activation techniques, if necessary, to increase the coating material-implant surface affinity. Finally, the coating material should be easily applied to implant surfaces with different geometries. While early homogeneous coatings could be applied to flat and even surfaces, nowadays, the heterogeneity of the coating is an important problem to overcome when it comes to the existence of titanium implants with three-dimensional printed personalized designs.

2.2.1 Organic Coatings

Deposition of nanofiber structures onto surfaces using the electrospinning technique represents a novel biomimetic approach to surface modification. The initial studies involving this pioneering technique, which seeks to achieve both nanoscale surface modification and biological activation, were first presented in 2011 [28]. This method enables imitation of the natural nanofibrillar collagen structures that exist within bone tissue. Silk fibroin protein, a commonly employed biopolymer in hard tissue applications, was investigated as a substitute for collagen in the modification of titanium surfaces. Along with silk fibroin nanofibers, hydroxyapatite-incorporated silk nanofibers were electrospun onto the surface to augment biomimicry [29].

Extracellular matrix-like surface modifications hold promise as methods to improve material biocompatibility. To hasten osseointegration on titanium, which is essentially a biopassive material, one method involves mimicking an extracellular matrix-like organization with differing materials in a structural or dimensional manner. Recently, there has been significant interest in immobilizing bioactive molecules on titanium surfaces for use in dental and orthopedic applications. Various techniques have been explored, including basic adsorption, silicate coupling, self-assembled monolayers, and sequential layer coatings [30–32]. It is important to note that these approaches should be evaluated objectively to determine their efficacy. Biodegradable polymeric thin film coatings have been reported to increase the biocompatibility of titanium surfaces without altering the metal's bulk structure. Multiple studies have shown them to be an effective modification [33, 34].

A successful implant material should properly interact with surrounding tissues and as many bone cells as possible, thereby triggering the differentiation of osteogenic stem cells. This phenomenon forms the basis of next-generation biomedical materials research [33]. In this context, the prioritized objective for surface modification research is the development of coatings able to mimic the extracellular matrix (ECM) of the metal surface. Biopolymers can be used as ECM analogs to provide this function to the metal surface, offering the possibility of a network structure capable of ready intercellular communication, biosignal transfer, and nutrient delivery on the implant surface post-implantation. This enables the implant surface to gain extra functions without compromising its mechanical bulk properties [35].

Biopolymers can be used as ECM analogs to provide this function to the metal surface, offering the possibility of a network structure capable of ready intercellular communication, biosignal transfer, and nutrient delivery on the implant surface post-implantation. Biomacromolecules including chitosan, poly lactide-co-glycolide (PLGA), heparin, and silk fibroin have been employed for surface functionalization, using techniques that do not compromise the metal's bulk structure. These techniques include electrophoretic deposition, electrospinning, and matrix-assisted laser vaporization [35–37]. Ravichandran et al. conducted a study that demonstrated an increase in initial cell adhesion and proliferation rates on surfaces coated with electrospun PLGA and PLGA/collagen nanofibers in comparison to control titanium surfaces. Following hydroxyapatite precipitation on both types of nanofiber, there were notable increases in alkaline phosphatase (ALP) activities of human mesenchymal stem cells

(MSC) compared to control and non-hydroxyapatite-doped fiber surfaces. Furthermore, it was observed that the cells were integrated into an ECM-like nanofiber network structure [36].

2.2.2 Inorganic Coatings

Hydroxyapatite-based surface modifications are a favored inorganic coating to expedite early osseointegration, thus enhancing implant success. Kokubo and his colleagues pioneered a technique that encourages growth of hydroxyapatite crystals on implant surfaces, immersed in simulated body fluid (SBF), resulting in a biomimetic layer facilitating the implant–tissue interface after the implantation process. Similar hydroxyapatite-coated implant surfaces can also be achieved through plasma spraying and electrophoretic deposition, an electrochemical process. SBF and electrophoretic deposition approaches involve the combination of ions in a liquid medium over time or through movement in an electric field, resulting in their accumulation on the titanium surface. Very small hydroxyapatite is physically coated onto the titanium surface at high temperature and speed using plasma spraying. Regardless of the deposition method employed for hydroxyapatite, a sintering process is typically conducted at high temperatures to enhance coating stability and consequently improve coating mechanical properties. Of course, these techniques have advantages and disadvantages over each other. Plasma spraying is a widely used and long-preferred method on an industrial scale, and it is the approach that requires the most expensive infrastructure, but it has the highest coating speed and is effective on complex shapes in terms of thickness and homogeneity. Electrophoretic deposition, on the other hand, requires a simpler mechanism and has a much slower coating speed than plasma spraying. Although soaking in simulated body fluid is a technique that is very sensitive to environmental conditions and takes a long time, it is a very promising approach in terms of bioactivity. In this technique, where crystal growth and thickness can be precisely controlled, it is possible for small molecules added to the liquid medium to be trapped in the gaps in the inorganic phase and to show activity on the implant surface or by being released [38].

3   Utilizing the Nanostructure

3.1   Enhanced Cellular Activity and Facilitated Osteogenic Differentiation on Nanostructured Titanium Surfaces

A bone implant should not only replace the tissues at the application site, but also integrate with living tissues to promote vascular regeneration and healing. In addition to promote new bone development and vascularization, the implant surface should also aid in soft tissue development, serving as a connecting link between the existing bone and the implant material in order to reinforce interaction and fixation. Insufficient bone cell proliferation on the implant surface is a leading cause of implant failure. This condition is necessary for effective bonding and integration with the surrounding bone tissue. Furthermore, debris accumulation from wear on the implant's protruding parts can result in nearby tissue cell death. Implant loosening and crack formation can result from load and tension imbalances between the implant and the tissue. Therefore, bone implants should facilitate bone regeneration and tissue formation or integrate quickly into nearby bone to mitigate these issues. The implant surface should be optimized to promote the favorable colonization of osteoblast cells and the synthesis of new bone. Successful implantations generally involve regenerated bone filling in the gaps between the implant and bone, leading to integration.

However, osseointegration may not always occur as planned because fibroblasts can occasionally favor the implant material over osteoblasts. This may lead to the development of undesired soft connective tissue on the implant surface. Since fibrous soft tissue differs significantly from hard bony tissue, fixation on the implant surface may not be uniform. This could potentially result in the implant becoming loose and failing during surgery. Excessive fibrous tissue formation indirectly impacts osteoblast and osteoclast activity, resulting in inhibited new bone formation. Therefore, the importance of materials that allow for early and rapid osseointegration is emphasized. Rapid bone tissue formation promotes implant–bone bonding while inhibiting the formation of unwanted soft tissue.

Early prediction and routine evaluation of an implant's success in a host relies upon the interactions between localized cells originating from injured tissue and the outer surface of the implant. These interactions can be divided into two distinct phenomena: adhesion of cells to the implant surface and proliferation of cells on it. After surgical implantation, the regeneration process at the interface involves hemostasis, inflammation, proliferation, and remodeling. During the hemostasis stage, the implant becomes covered by fibrins, blood vessels, and blood proteins, and is defined as a foreign body.

During the inflammation stage, macrophages are recruited by the immune system to eliminate foreign bodies through phagocytosis, or collagen fibers are deposited to induce differentiation of MSCs into local tissue cells, promoting wound healing. Before attachment of osteoblasts or other cells to the surface of the implant, they interact with proteins present in bodily fluids like bone marrow and blood. This leads to initial protein adsorption on the implant surface which subsequently regulates cell adhesion [39]. At this stage, specific molecules and molecular zones found on the cell membrane of newly differentiated cells cause the cells to adhere to the implant surface and extracellular matrix, such as fibronectin and its arginine-glycine-aspartic acid (RGD) sequence [40, 41].

Several studies have shown that free energy, chemical composition, and surface properties affect protein adsorption. For example, fibronectin adsorbs more on a calcium phosphate-coated bioactive glass surface than on an uncoated one [42]. Additionally, it is acknowledged that osteoblasts have a preference for attaching to particular RGD sequences and heparin sulfate-binding sites present in adsorbed proteins [43]. It is acknowledged that the interaction between specific sites in surface-attached proteins and integrin receptors on cell membranes plays a critical role in cell adhesion. As such, bone formation efficiency relies solely on the surface qualities, including physical attributes and chemical composition, of the implant. These traits also moderate the adhesion of both necessary (osteoblast) and unwanted (fibroblast) cells. Altering the surface chemistry and topography of a hard tissue implant is considered as the primary approach to develop intelligent implants through the manipulation of protein and cell interactions.

Although the interaction between membrane proteins and the implant occurs between the first layer of cells and the implant, the majority of the newly formed tissue is generated by the proliferation of cells. These cells communicate with their neighboring cells through signaling pathways and have the potential to compromise the mechanical stability of the first cell layer due to physiological forces. Attached cells attract the nearest undifferentiated MSCs, mechanically binding them together with tight junctions. This process provides information about the environment and identifies the type of cell required to differentiate and fill the physiological space. At the end of the remodeling stage, complete differentiation occurs, and tissue formation concludes. Figure 1 depicts cellular adhesion and integrin–surface interactions schematically [44]. Unless the attraction between fibronectins and the implant surface is sufficient to spread the cell and generate tension within the actin filaments, which construct the endoskeleton of the cell, poor attachment can occur and lead to detachment of a differentiated cell, resulting in cell death and reduced implant success. To prevent cell death following implantation of a metal biomaterial, it is essential to consider how surface properties like rigidity, roughness, charge, chemical functionalities, topography, and wettability can affect cell response. It is important to understand these factors to improve outcomes of implantation procedures [45–50].

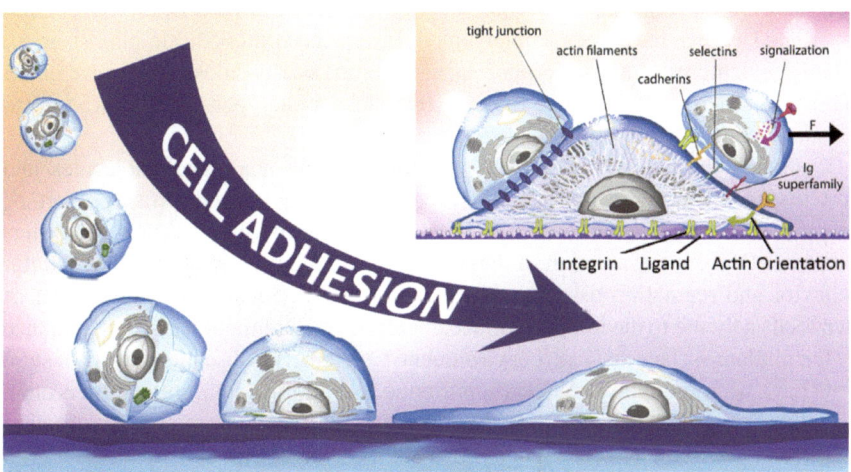

Fig. 1 Schematic expression of cell adhesion and biomechanical interaction between cells and an implant surface [44]

As previously noted, achieving ideal surface properties is crucial for inducing new bone formation in orthopedic implants employing nanophase materials. The interaction between the surface and initial proteins plays a vital role since changes to surface properties affect cellular activity. Nanophase materials are thought to interact more effectively with proteins than traditional materials, potentially enhancing osteoblast functionality. Pioneer researches with nanophase materials indicate that nanophase materials exhibit higher protein bioactivity than conventional materials [51, 52]. The following studies also indicated that particular proteins, specifically fibronectin and vitronectin, display significantly enhanced surface adhesion and conformation (bioactivity) when interacting with nanophase materials [53, 54]. Nanofibrous scaffolds fabricated from the same polymeric material (poly-L-lactic acid) have been reported to absorb greater quantities of fibronectin or vitronectin, than a monolithic membrane [55, 56]. An increase in protein adsorption was observed with surface area enlargement and material transport. Furthermore, the altered surface energy of nanophase materials enhances protein interaction.

While the increasing rate of adhesion proliferation of the cells has a significant impact on the success of biomaterial implantation, it is also crucial for stem cells to differentiate into appropriate cells during new tissue formation. Since most tissues consist of multiple cell types, such as osteoblasts, osteoclasts, and osteocytes in bones, proper control and guidance for cell differentiation is necessary.

Researchers have explored altering surface roughness to facilitate osteogenic cell differentiation. Reports suggest that creating micron-scale (<10 μm) roughness on the surface of titanium implants through methods such as etching, sandblasting, or coating with micrometer-sized particles can improve osteoblast functions. Such functions include adhesion, proliferation, alkaline phosphatase production, and the calcium deposition [57–59]. Several surface modification processes mentioned earlier have been implemented within the orthopedic implant market. Titanium, demonstrating a microporous surface, manifests an osteoinductive effect apart from microporosity, while titanium that lacks a microporous structure fails to stimulate bone formation [60, 61].

Cellular communication between neighboring cells via signal molecules and proteins, as well as microenvironmental factors, including implant surface properties, extracellular matrix, small molecules, ions, and other variables, impacts cell differentiation. Objective evaluation of these factors is crucial for understanding cellular behavior and regulating tissue engineering [62, 63]. Typically, the initial layer of stem cells adhered to the implant surface undergoes differentiation through guidance by the implant surface and microenvironment. Conversely, subsequent layers of stem cells have been indirectly guided via hormone reception and differentiation protein signaling from differentiated cells. Anticipatedly, extracellular matrix construction by pioneer differentiated cells guides other stem cell layers, along with the structural properties of the matrix-like stiffness or topography. Although metallic biomaterials, such as stents or dental screws, have been developed, the surface properties must be engineered to not only allow for cellular adhesion but also direct stem cells to differentiate into suitable cells [64–67]. Forces driving stem cell differentiation on an implant surface result from tension on the actin filaments, causing cytoskeletal

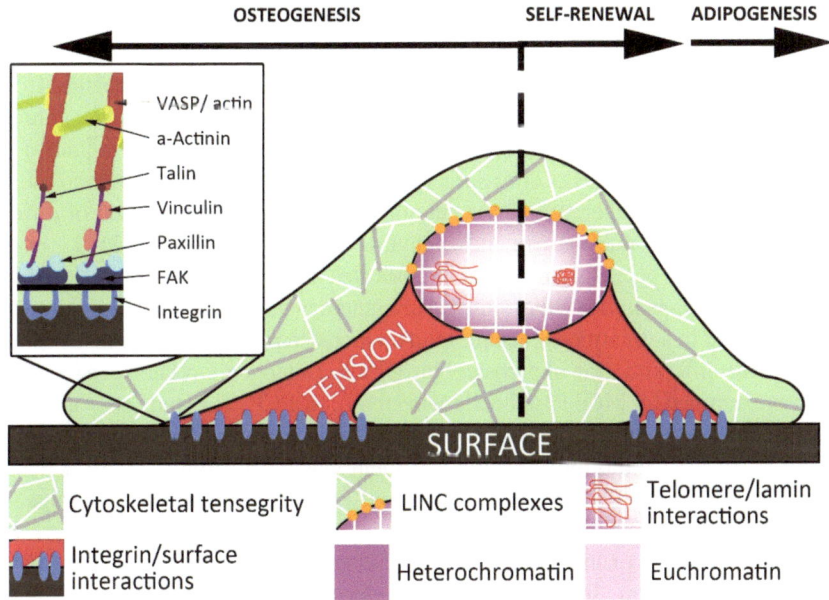

Fig. 2 Schematic expression of the biomechanical surface–cell interactions to drive a stem cell to differentiate a bone cell

tensegrity. This can alter both cytoplasmic orientation and intranuclear telomere–lamin coordination for differentiation through mechanotransductive signals. Changes in the nucleoskeleton alter territorial chromosome positions and modify genome transcription during cell division [68]. Differentiation occurs after the cell division of a stem cell, with its cell fate predetermined by the surface properties. It is important to note that the exact type of differentiated cell is determined by overall biochemical and biomechanical cues. The described mechanism is illustrated in Fig. 2.

It should be noted, however, that the degree of roughness typically favored by osteoblasts in the body exceeds the levels of roughness achieved through surface modifications. Since bone tissue contains nanostructured components, it is essential to produce surfaces with both nanometer and micron roughness. Such designs will improve the effectiveness of any material-based implants. In the nanoscale structure of bone tissue, the organic (collagen fibrils) and inorganic (bone minerals) constituents are natural nanomaterials. Creating materials that replicate the composition and microstructure of natural bone is a viable approach to enhancing orthopedic materials. The utilization of nanophase materials in implant design is becoming increasingly popular due to the similarities in biological scales and the superior effectiveness of nano-roughness on bone prostheses when compared to micro-roughness. Studies have extensively examined this effect, with evidence indicating faster osseointegration on surfaces exhibiting nano-roughness than those displaying conventional or micro-roughness. Studies on the importance of nanostructures in bone proliferation and osteogenic differentiation are still ongoing today. Chen et al.

demonstrated that titanium dental implants, which underwent ion implantation treatment with oxygen plasma immersion, can be activated for cell differentiation and osseointegration through the alteration of titanium dioxide's crystal structure from anatase to rutile without any physical surface changes. Strong bone–implant contact was reported during in vivo studies of plasma-treated dental screws implanted in the femurs of rabbits 4 weeks after surgery. In vitro studies indicated an increase in ALP activity and stronger von Kossa staining, suggesting the promotion of differentiation in human bone marrow mesenchymal stem cells [69]. In 2021, Le et al. published a study on titanium implants fabricated through selective laser melting (SLM). The findings indicate that sintered implants can be bioactivated through additional acid and heat treatments, resulting in an increase in surface roughness, which enhances the differentiation of preosteoblastic cells. Osteogenic-related gene expression was examined for MC3T3-E1 cells using real-time PCR. The study also observed cell attachment on additively manufactured implants, as reported in Fig. 3. Furthermore, the success of the bioactivation treatment was demonstrated by an increase in the expression of ALP, OCN, RUNX2, and OPN genes, known as osteogenic differentiation factors [70].

Li et al. recently examined interactions between bone-marrow-derived MSCs and different titanium dioxide nanotopographies on titanium implants, namely, nanowires, nanonests, and nanoflakes. The researchers systematically analyzed surface area, roughness, hydrophilicity, surface zeta potential, and other nanomaterial-originated features to determine the optimal nanotopography using in vitro and in vivo studies. Although all nanotopographies were achieved through similar hydrothermal seed-growth and plasma-spray mechanisms on the same implants, they exhibited different behaviors in monoculture and co-culture experiments. As depicted in Fig. 4, all nanotopographies exhibited more hydrophilic behavior than pristine titanium. However, titanium dioxide nanonests displayed optimal surface roughness, surface area, and zeta potential in physiological pH, resulting in less accumulation of M2 macrophages and avoiding undesired inflammatory response than M1 and M0 macrophages that could cross talk to bone-marrow-derived MSCs and lead to better osseointegration through differentiation [71].

Titanium dioxide nanotube arrays have been extensively researched as a surface engineering method for titanium implants due to their biocompatibility, cost-efficient fabrication, high homogeneity, and the ability to achieve high surface coverage and geometric control of morphology, thanks to the use of titanium dioxide nanoparticles, nanowires, nanonests, and treated natural oxide in the implant. The growth mechanism for nanotubes has been well understood for years. Following the implantation process of a titania nanotube array-decorated implant, the surface was immediately hydrated by water molecules. This hydration may be necessary to allow proteins to immobilize in an available conformation due to the high surface area and charge. Simultaneously with the immobilization of the osteoblast attachment-inducing protein, vitronectin, on the surface, hydrophobic proteins, namely, serum albumin and IgG2, can also be immobilized on the surface at less hydrophilic sites and impurities. The hydrophilic nature of the nanotubular topography plays a crucial

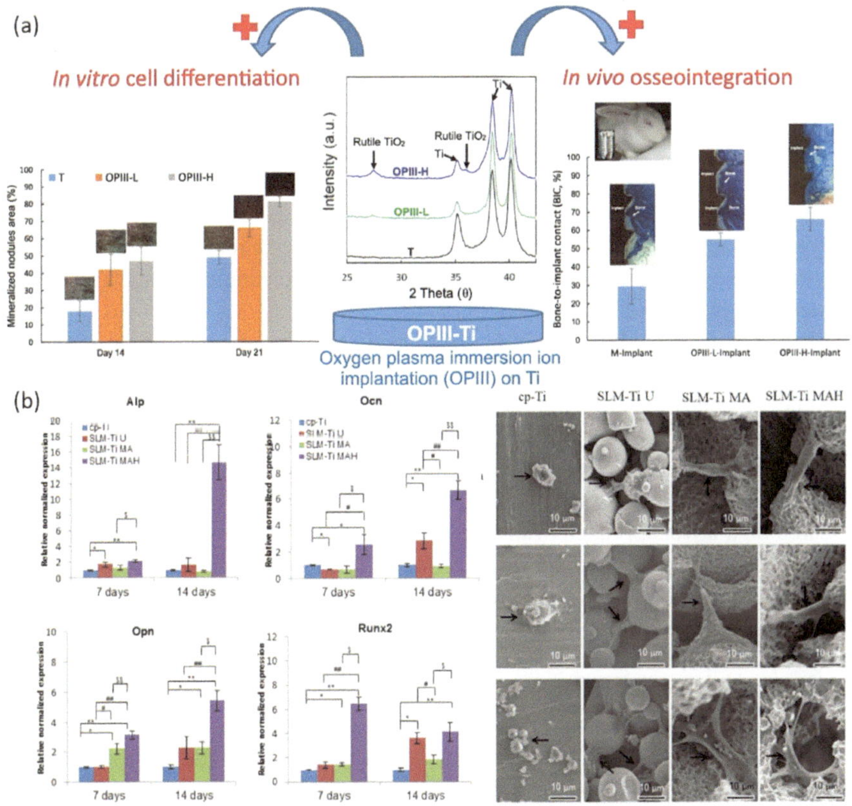

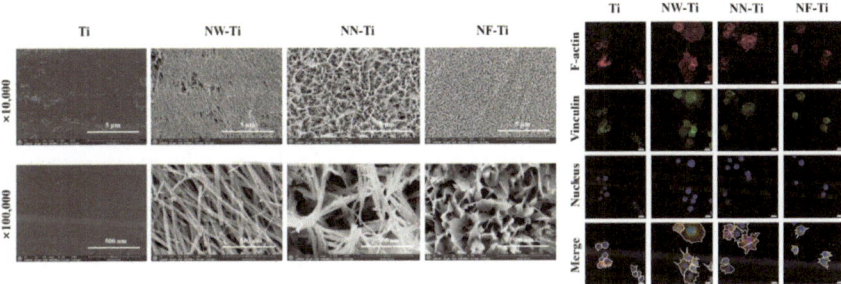

Fig. 3 Altering oxide layer crystal structure by oxygen ion plasma treatment to promote cell differentiation and osseointegration (**a**) [69] and altering surface roughness by mixed acids and heat treatment to promote cell differentiation and osseointegration (**b**) [70]

Fig. 4 Surface feature comparison between nanotopographies and confocal microscopy images after culture of RAW264.7 machrophages on different nanoengineered titanium surfaces [71]

role in preventing unexpected foreign body reactions, as IgG2 tends to express pro-inflammatory cytokines as M1 and M0 macrophages, while serum albumin expresses anti-inflammatory cytokines as M2 macrophages. This effect is known as macrophage polarization. Cells adhere to the surface in accordance with prior descriptions. Expression of differentiation and osteogenic genes, as well as growth factors and cross talk, have been more dominant on nanotubular surfaces than on flat surfaces. This is because nanotubular spacing and geometries can cause more stretching in lamellipodia and filopodia than untreated titanium surfaces. Additionally, macrophage polarization and expressed genes inhibit angiogenesis. The results are depicted in Fig. 5 [72].

Most metals, including titanium, are unavailable due to electrostatic repulsion between metal atoms and positively charged calcium ions. However, phosphorus has a negative charge, allowing for naturally crystallized hydroxyapatite formation. The bone–implant contact is crucial for the mechanical stability of an implant in a host, leading to numerous studies on the mineralization of synthetic hydroxyapatite crystals for bone and implant surface formation. Wang et al. published their research on a novel approach to surface decorating titanium implants using negative pressure immersion of calcium and phosphorus and hydrothermal crystallization of hydroxyapatite. The study aimed to enhance osseointegration by promoting bone mineralization through the interaction between hydroxyapatite in titania nanotube arrays synthesized artificially and newly formed bone. Following in vitro investigations using MC3T3-E1 cells, the surface resulting from the synthetic in situ crystallization process was determined to be non-toxic and exhibited higher ALP activity. Additionally, both bare titanium and unloaded titania nanotubular surfaces demonstrated a slower proliferation rate with decreased filopodia and lamellipodia texture compared to the newly developed surface. After conducting in vivo research on implantation at rat femurs, both the micro-CT and histological analyses indicated that hydroxyapatite-loaded nanotubes had better bone–implant contact than unloaded and pristine titanium with contact gaps [73].

3.2 Antibacterial Properties of Nanostructured Titanium Surfaces

Problems including bacterial infections, inflammations, and bone loss may occur post-implant surgery [74, 75]. Peri-implantitis infections are a common cause of implant failure [76, 77]. After implant surgery, the chance of infection ranges from 0.5 to 5%, with a 14% risk of reinfection after revision surgery and an estimated cost per patient of $50,000 USD after infection, with a mortality rate of approximately 2.5% [78–80].

Prosthetic infections are generally result of biofilm forming bacterial invasions [81]. Despite meticulous sterilization and aseptic techniques, microorganisms can still form biofilm on the surface of the implant during the operation or post-op time.

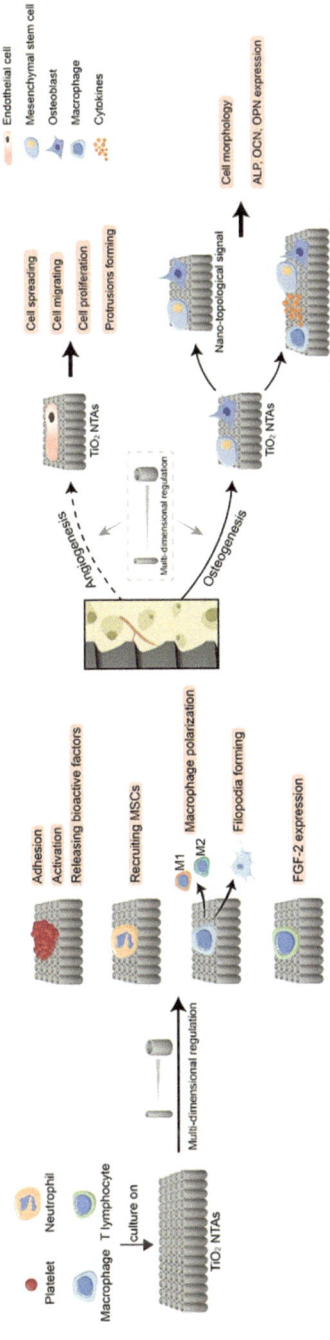

Fig. 5 Schematic expression of titania nanotube array modification and its physiological pathway through osseointegration [72]

Furthermore, it has been observed that bacteria which form biofilms exhibit resistance to immune cells and antibiotics up to 1000 times greater when compared to bacteria that do not form biofilms [82–84].

The development of surface technologies that combat industrial and health issues stemming from bacterial infections is a pressing challenge. Therefore, addressing this challenge is crucial for effectively combating bacterial infections. Achieving antibacterial surfaces necessitates the development of methods to prevent biofilm formation, which occurs when microorganisms adhere to various surfaces and establish a resilient and defensive microorganism community [85]. Researchers commonly use materials with antibacterial properties, including metals, semiconductors, polymers, and nanocomposites, to address bacterial infections [86, 87].

Surfaces made of nanomaterials have received significant attention for their innovative solutions to persistent bacterial infections. The increased curiosity in nanostructured materials arises from their remarkable ability to impede bacterial adhesion on a mechanical level, which ultimately decreases the necessity for disinfecting agents with chemicals [86, 88]. Nanostructured surfaces possess nano-sized pillars and textures that function as robust barriers against bacterial adhesion. These characteristics create significant challenges for bacteria in establishing initial attachment points, resulting in mechanical disruption that reduces bacterial colonization and formation of biofilms [88, 89].

The field of nanostructured materials provides numerous options, among which titanium has garnered attention in combating bacterial infections. Titanium offers biocompatibility, strength, durability, and inherent corrosion resistance in both medical and industrial settings [83]. Furthermore, its ability to be engineered into nanostructured surfaces makes it an ideal candidate for tackling the challenges posed by bacterial infections. Nanostructured titanium surfaces are commonly used in medical implants, medical devices, wound dressing design, and industrial applications to prevent biofilm formation [89, 90]. Titanium-based materials may be vulnerable to implant-associated infections caused by bacteria such as Staphylococcus aureus (S. aureus), Staphylococcus epidermidis (S. epidermidis), Pseudomonas aeruginosa (P. aeruginosa), and Escherichia coli (E. coli). These bacteria possess a propensity to adhere to the implant surface, colonize it, and generate a biofilm, resulting in complications such as chronic inflammatory responses, tissue destruction, and loosening of the bone–implant interface [91].

To tackle these challenges, several approaches have been adopted, such as altering implant surfaces through the inclusion of nanoparticles, peptides, oxidants, and drug agents with antibacterial properties. Nevertheless, these methods frequently exhibit restricted efficacy and have the potential to damage adjacent healthy cells. To address these issues, promising solutions involve the re-modification of titanium alloy surfaces at the nanostructure level and the imitation of natural antibacterial nanostructures discovered in the skins of geckos, cicadas, and dragonflies [89, 91].

In the study conducted by Bright and colleagues, the objective was to engineer a frequently used implant material, Ti-6Al-4V alloy, with a specialized nanoarchitecture inspired by the structure of a dragonfly wing. The selected unique nanoarchitecture was chosen for its mechanical ability to disrupt bacterial membranes due to its high aspect ratio, where the length of the nanostructures greatly exceeds their width. The study examined the effects of this engineered material on the viability of P. aeruginosa and S. aureus bacteria, and found a significant decrease in their viability [89]. Additionally, TiO_2 nanostructures, with their photocatalytic properties that allow for the degradation of microorganisms through light-triggered redox reactions, are utilized in antibacterial applications alongside titanium alloys [92–94].

For a comprehensive understanding of the antibacterial properties of nanostructured TiO_2 surfaces, it is necessary to have an in-depth understanding of their photocatalytic mechanisms. Titanium dioxide nanostructures function as semiconductors, absorbing UV light and creating electron–hole pairs. Excited electrons transition from the valence band to the conduction band, creating positively charged holes. Subsequently, electrons and holes interact with ambient oxygen, water, and organic compounds on the TiO_2 surface, resulting in decomposition. As shown in Fig. 6, holes oxidize water or organic compounds to produce hydroxyl radicals (•OH), whereas electrons react with oxygen, producing superoxide ions ($O_2 \cdot^-$) and other reactive oxygen species (ROS). The reactive substances produced can lead to the oxidation and degradation of phospholipids in bacterial cell membranes, eventually resulting in membrane disruption and the death of bacterial cells [92, 93, 95].

The study by He et al. study demonstrated the potential of TiO_2 as an antibacterial application in orthopedic and implant procedures due to its photocatalytic activity. The researchers modified the implant surface using TiO_2 infused with carbon dots, resulting in a hybrid surface with additional photodynamic therapy properties when exposed to UV light, as well visible light and near-infrared light. The improved feature results in significant antibacterial activity exhibited by the surface modification, effectively eradicating bacteria like Staphylococcus aureus [96].

Black TiO_2 nanostructures have been recently introduced in literature to address the concern of photocatalytic properties under UV light caused by high band gap. It is recognized that alternative methods, including TiO_2 crystal lattice ion loading

Fig. 6 Illustration of the photocatalytic mechanism of TiO_2 nanostructures and visualizing the mechanism of bacterial inactivation through photocatalytic reactions

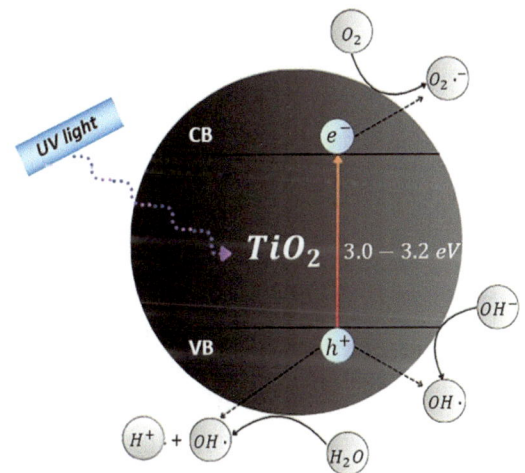

or surface modification mentioned in prior research, may encounter challenges in improving photocatalytic efficiency to the desired level. In contrast, the objective of black TiO_2 is to induce an oxygen vacancy (V_o) in the titanium crystal lattice, thereby enhancing light absorption capacity by decreasing the band gap. Additionally, it was found that V_o contributes to the production of ROS by functioning as both an electron acceptor and donor, resulting in the activation of extracellular electron transfer (EET) in bacteria on black TiO_2 nanostructures [97].

3.3 Nanostructured Titanium Surfaces as Localized Drug Delivery Devices

To avoid potential harm caused by traditional drug therapy methods and considering the inadequacy of oral or intravenous routes in achieving a sufficient dose of efficacy in local infections, there has been increased interest in implant-centered drug release methods. The use of nanostructured titanium surfaces in drug delivery is a significant step toward achieving more effective and personalized therapies. The nanoscale structure of titanium lends it diverse chemical and physical properties, making it a controlled-release platform with greater sensitivity than traditional methods [98]. This approach eradicates the drawbacks linked to conventional drug delivery systems, including poor bioavailability, limited duration of action, drug degradation, and a lack of spatial and temporal control [99]. The mitigation of these drawbacks is essentially facilitated by tailored control through the surface modifications of the titanium surface. These modifications, which include nanotexturing, coatings, and functionalization with specific properties, have a crucial impact on determining both the release kinetics and site specificity of the drugs they transport [100].

The modifications can be categorized as follows:

- Structural modifications that alter the diameter and length of titanium nanostructures.
- Surface modifications that alter the hydrophilic, hydrophobic, and charge state of the surface.
- Adjustments can be made to the pores that release the drug by coating the surfaces of the drug carrier with a tubular nanostructure composed of biodegradable polymers like PLGA.
- Additionally, modifications can be made by incorporating drug carrier systems such as micelles or dendrimers into the surface of the nanostructure.
- Modifications can be made to stimulate external factors including temperature, magnetic fields, ultrasound, light, and radiofrequency. Light stimulation proves to be highly effective in TiO_2 nanostructures due to their photocatalytic properties.

Modified strategies for drug delivery, tailored toward the structure of titania nanotubes (TNTs), are illustrated in Fig. 7.

Fig. 7 Strategies for controlling drug release from TNTs. **a** Controlling nanotube diameters and length; **b** Manipulating surface chemistry (hydrophilic, e.g., APTES/3-Aminopropyltriethoxysilane—due to containing a silane group); **c** Degradation of dip-coated polymer films (PLGA or chitosan); **d** Using drug nanocarriers (micelles); **e** External field-triggered drug release, specifically light (modified from [98])

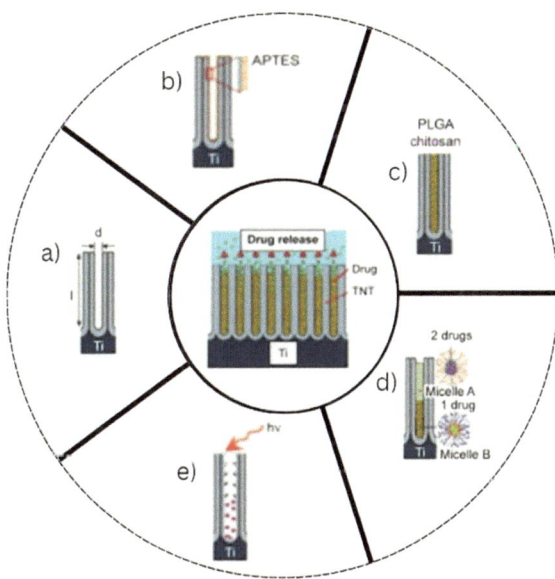

The primary application of titanium nanostructures as drug delivery mechanisms is in the field of orthopedics, where titanium implants serve as the primary material. In this domain, nanostructured titanium surfaces incorporated into implants are subjected to modification with antibacterial or anti-inflammatory substances, which are subsequently released at targeted surgical sites in a controlled manner. Alongside an accelerated recovery post-surgery, this process has the added benefit of reducing the likelihood of infections [100].

The use of nanostructured titanium surfaces extends beyond orthopedics to medical specialties like dentistry and oncology. In dentistry, titanium nanostructures provide a unique approach to combat oral infections. Controlled and sustained release of therapeutic agents increases their effectiveness against oral pathogens while minimizing the need for frequent application, leading to more efficient treatment.

Fathi et al. conducted a study on a unique approach to designing orthopedic implants by incorporating TNTs possessing antibacterial properties and high biocompatibility. To strengthen osseointegration, TNT structures were anodized electrochemically and utilized to coat a titanium implant. The silk fibroin nanofibers were electrospun to create a coating for controlled drug release. Vancomycin was loaded as the primary clinical drug to treat severe infections. Silk fibroin nanofibers controlled drug release and affected proliferation rate of osteoblasts. The study proved that Vancomycin-contained TNTs are found effective surfaces for reducing the number of adhesion, particularly against Staphylococcus aureus. The study findings suggest that this approach has potential for preventing implant infections and improving osseointegration in orthopedic applications [101]. Specifically, the study demonstrated non-toxicity to cells and enhancements in drug release characteristics. In a similar approach, Keceli et al. tried to obtain a sequential release profile by coating BMP6-loaded TiO_2NT structures with PDGF-containing silk fibroin nanofibers [102]. The proposed system offered considerable potential for enhancing early osseointegration duration via an improved factor release curve and by contributing to osteoblastic cell proliferation, mineralization, and differentiation.

Interest in drug-loaded titanium surfaces obtained without polymers has notably increased in the past 5 years. Drugs can be loaded into nanotubular cavities obtained after anodization, resulting in release profiles ranging from several hours to several weeks [103, 104]. Parameters that influence the release times include the hydrodynamic diameter of the loaded biomolecule, nanotube length and diameter, and differences in loading procedures. Popat and colleagues reported a 40% reduction in S. epidermis colonization on the nanotubular layer filled with gentamicin using a simple pipetting approach compared to the control group. Coating nanotubular surfaces with polymeric or ceramic layers during loading procedures can result in drug release times of several weeks. Another study reported that the release process from the metal surface without an additional surface coating concludes within a few days [105, 106].

Although research indicates that polymeric or ceramic coatings can enhance release duration by expanding and maintaining compatibility with biological systems in vitro, the risk of micro-thick coating detachment or breakdown during implantation or under dynamic conditions still exists. In drug loading studies without barrier layers, the length of the nanotubular structure is the key parameter influencing release time, while the nanotube diameter (when length is constant) is the most crucial factor affecting the drug load capacity. A previous study conducted by Peng et al. provides evidence of this [107].

Several studies have assessed the use of TNT surfaces for local drug release, taking into account the evidence of their in vitro and in vivo biocompatibility. Various techniques have been proposed to achieve effective drug concentration and sustained drug release. Addressing the primary barrier of drug release within the first few hours following implantation is essential. Additionally, achieving zero-order release kinetics is a significant objective to ensure uniform drug release throughout each time unit interval. Zero-order drug release kinetics are the preferred method in pharmaceutical studies. This approach guarantees a uniform release of drug per unit time, irrespective of the concentration or time. However, varying factors like diverse drug target sites and illness-specific treatment demands can lead to divergent drug release profiles. There is a need for drug release systems that can provide singular or multiple release times, lasting longer or shorter, and activated on demand or at specific time intervals. As a result, TNTs' adaptable drug release approaches may offer a viable solution for various problems.

The most straightforward approach to achieving kinetic control of drug delivery systems based on TNT is by modifying the dimensions of nanopores. By altering the diameter and length of the two structural components of TNTs during anodic oxidation, we can modify them. The release of drug molecules from nanotubes into the environment is determined by the size of the nanoscale structures through which they are released, as dictated by Fick's law of diffusion. There have been studies in the literature that have observed changes in drug release kinetics or duration by means of this approach [105, 106, 108, 109]. Through these studies, the release of small and large molecular drugs, such as paclitaxel, gentamicin, indometacin, or albumin, can be prolonged by altering the diameter and length of nanotubes. During drug release, the drug will be loaded deeper into the nanotubular voids, causing capillary effects and resulting in a longer time for the nanotubular void to be evacuated. However, reducing the nanotube width will also affect the amount of drug that can be loaded into the nano-reservoir, resulting in a lower drug storage capacity. In another study examining the impact of aspect ratio on drug release profiles, researchers reported that the release time for gentamicin antibiotic increased as the aspect ratio increased. The same study found that as the nanotube diameter increased, the loading amount also increased between the tubes. Researchers demonstrated that gentamicin-loaded TNT surfaces provide effective protection against S. aureus bacteria compared to other studies that examined the surfaces [109]. All of these studies hold promise, but the use of size-controlled modifications for long-term emissions on the moon is not feasible. Functionalization of nanotube surfaces is highlighted as another method in drug delivery strategies. The aim of this approach is to add hydrophilic and hydrophobic properties to TNT surface chemistry. This creates a dynamic equilibrium between the drug-nanotube wall, enabling alteration of drug loading and release kinetics. This approach can be achieved on Ti surfaces through spontaneously ordered monolayers or polymers. The functional monolayers produced are highly stable and offer various options for functional groups, charge, and binding properties. Approaches using 3-aminopropyl triethoxysilane (APTES) and phosphonic acids have been investigated for their loading capacity and release profiles of hydrophilic and hydrophobic drugs, but this strategy does not have sufficient potential for long-term sustained release [110].

The use of polymers as a barrier layer against the diffusion of drugs and similar molecules in tube openings offers advantages in various areas, such as diverse biocompatible polymer varieties, extracellular matrix interaction, functionalization, and control of cell interactions. Due to the nanometer-scale dimensions of the nanotubes, precise control of the polymer coating process must be maintained. Plasma polymerization emerges as a precise approach that can provide this barrier in nanotube openings [111]. Through the shrinking that occurs at these openings, improvements can be made in drug delivery by increasing release time, approaching zero-order kinetics, and enhancing tissue integration angles, as previously discussed. While it is possible to achieve polymer coatings with different properties using plasma polymerization, there is also a wide variety of drugs that can be released. Studies in the literature have investigated small-molecule drugs such as anti-inflammatory, antibiotic, and anticancer drugs, as well as large-molecule structures including bovine serum albumin, growth hormone, and bone morphogenetic protein, which have been located in studies where tube openings were modified using plasma polymerization [111, 112].

An alternative method to extend and regulate the duration of drug release is to employ drug carrier nanosystems. Implementing polymeric micelles or capsule-like nanostructures within nanotube reservoirs can yield various advantages in managing drug delivery controlled by TNT. Polymeric carriers could potentially safeguard large molecules, such as protein drugs and oligonucleotide sequences, from external factors like pH or enzymes within the environment. Alternatively, this technique could be used to extend release time by creating an additional barrier. Published studies using polyethylene glycol (PEG)-based polymeric carriers have demonstrated drug release profiles lasting several weeks from TNT surfaces, without sudden initial release [113]. In addition to extending the release profile, drugs loaded into polymeric micelles that are immiscible with each other (e.g., hydrophilic and hydrophobic) can also be sequentially delivered. This enables the sequential release of molecules, such as anti-inflammatory and antibiotics, at different time periods [114].

Most local drug-releasing implant approaches based on TNT have focused on prolonged and optimized drug release of high drug concentrations throughout the treatment course. However, there may be various active molecules required at the implantation site during different stages of treatment due to momentary changes in the advancing scenario. Examples of complications include bacterial invasion, inflammation, osteomyelitis, and septic arthritis. To address these emergencies, drug transport/distribution systems triggered externally are recommended. Numerous externally triggered systems have been researched to increase desired therapeutic efficacy within the required timeframe, administering high doses or treating pathogenic invasion over a specific time period. Several methods have been reported for actuation, including thermal and magnetic triggering [115–117].

Many studies have investigated temperature-sensitive polymers for externally triggered TNT surface systems that release drugs with different nanocarriers. As an example, Cai et al. used Vitamin B2 as a model drug to evaluate its efficacy in controlling the release of drugs from TNTs [115]. A hydrogel-based coating was applied to the surface of TNT sheets in this study. The hydrogel was synthesized using a temperature-sensitive polymer, poly N-isopropyl acrylamide (PNIPAAm), with reversible shape change at lower critical solution temperature (LCST). The hydrogel possesses a water-swelling state at lower temperatures, while at high temperatures the structure collapses to trigger the active molecules to be released from TNTs. Given that body temperature increases to above 38 °C during inflammation, the onset of inflammation can serve as a temperature-controlled mechanism for triggering localized drug release therapy.

Magnetic nanoencapsulation and magnetic field-triggered drug release have received considerable attention in developing local drug delivery devices. This approach has been applied to TNT-based systems by Shrestha et al. to develop a magnetic and photocatalytic-guided drug release system using TNTs filled with magnetic nanoparticles (MNPs) [116]. Another strategy proposed by Aw et al. involves MNP-loaded TNTs triggered by internal and external magnetic fields [117]. Drug release from this TNT-based system occurs through the interaction between the MNPs at the bottom with an external magnetic field, which forces the release of the drug-loaded micelles.

Exciting and promising studies using TNTs and external triggers for drug release have been reported in recent years. These sophisticated approaches can be used to plan the production of smart drug-releasing implants, which undoubtedly have significant potential in this field. In addition, these kind of methods are also anticipated to be used in various applications, such as bone treatment and combination with local or systemic chemotherapy.

Another study has been reported for the differences of inflammatory responses between TNTs and flat titanium surfaces [118]. Markers of immune response including monocytes, inflammatory cytokines, and reactive oxygen species were evaluated. Results indicated that compared to Ti surfaces without nanotubes, TNT arrays significantly reduced the inflammatory response. Several studies demonstrating the potential efficacy of TNT reservoirs for the controlled release of indomethacin were conducted by Losic et al. Throughout the studies, experiments samples were used in both flat and wire forms [119].

4   Concluding Remarks

As previously mentioned, the main goal of the studies is to create implant materials that are receptive to cellular processes and provoke positive responses from osteoblast cells, encompassing initial cell attachment, growth, and transition from non-calcium-storing cells to calcium-storing cells. Additionally, it is crucial for both osteoblast and osteoclast cellular activities to synchronize to maintain the overall health of the surrounding bone. Due to significant advancements in cell biology, the design and production of orthopedic implants have transitioned from a trial-and-error approach to the creation of biomaterial surfaces that are cell-friendly and can induce the intended biological response.

Studies published over the past decade indicate that surfaces composed of TNTs are a highly promising biomaterial for local drug delivery systems that can overcome the drawbacks of conventional treatments. TNTs can be created using cost-effective electrochemical processes scaled up for production. Furthermore, biocompatible materials used in medical implants for many years can be utilized as drug delivery implants. TNT surfaces can be created on different medical implants, including titanium hip prostheses, stents, fixators, screws, and plates. The high surface area, controlled nanotube sizes, modifiable surface chemistry, and the tunable release kinetics of TNTs suggest that they may be highly effective drug delivery vehicles. Furthermore, TNT surfaces are not considered as biocompatible materials only both also exhibit a resistant character against chemicals and corrosive materials, as well as offering a mechanical and thermal stability. As a result, it is considered a biomaterial with bioactive properties, particularly in mineralization; osseointegration; and, most importantly, the facilitation of bone cell growth, adhesion, differentiation, and proliferation.

References

1. Marino CEB, de Oliveira EM, Rocha-Filho RC, Biaggio SR (2001) On the stability of thin-anodic-oxide films of titanium in acid phosphoric media. Corros Sci 43:1465–1476. https://doi.org/10.1016/S0010-938X(00)00162-1
2. Bandyopadhyay A, Mitra I, Goodman SB et al (2023) Improving biocompatibility for next generation of metallic implants. Prog Mater Sci 133:101053. https://doi.org/10.1016/j.pmatsci.2022.101053
3. Yamada K, Ito M, Akazawa T et al (2015) A preclinical large animal study on a novel intervertebral fusion cage covered with high porosity titanium sheets with a triple pore structure used for spinal fusion. Eur Spine J 24:2530–2537. https://doi.org/10.1007/s00586-015-4047-2
4. Bjursten LM, Rasmusson L, Oh S et al (2010) Titanium dioxide nanotubes enhance bone bonding in vivo. J Biomed Mater Res A 92A:1218–1224. https://doi.org/10.1002/jbm.a.32463
5. Larsson C, Thomsen P, Aronsson B-O et al (1996) Bone response to surface-modified titanium implants: studies on the early tissue response to machined and electropolished implants with different oxide thicknesses. Biomaterials 17:605–616. https://doi.org/10.1016/0142-9612(96)88711-4
6. Kim HM, Miyaji F, Kokubo T, Nakamura T (1997) Effect of heat treatment on apatite-forming ability of Ti metal induced by alkali treatment. J Mater Sci Mater Med 8:341–347. https://doi.org/10.1023/A:1018524731409
7. Kokubo T, Kim H-M, Kawashita M, Nakamura T (2004) Review bioactive metals: preparation and properties. J Mater Sci Mater Med 15:99–107. https://doi.org/10.1023/B:JMSM.0000011809.36275.0c
8. Sittig C, Textor M, Spencer ND et al (1999) Surface characterization of implant materials c.p. Ti, Ti-6Al-7Nb and Ti-6Al-4V with different pretreatments. J Mater Sci Mater Med 10:35–46. https://doi.org/10.1023/A:1008840026907
9. Sato M, Slamovich EB, Webster TJ (2005) Enhanced osteoblast adhesion on hydrothermally treated hydroxyapatite/titania/poly(lactide-co-glycolide) sol–gel titanium coatings. Biomaterials 26:1349–1357. https://doi.org/10.1016/j.biomaterials.2004.04.044
10. Baker KC, Anderson MA, Oehlke SA et al (2006) Growth, characterization and biocompatibility of bone-like calcium phosphate layers biomimetically deposited on metallic substrata. Mater Sci Eng, C 26:1351–1360. https://doi.org/10.1016/j.msec.2005.08.015
11. Furlong R, Osborn J (1991) Fixation of hip prostheses by hydroxyapatite ceramic coatings. J Bone Joint Surg Br 73-B:741–745. https://doi.org/10.1302/0301-620X.73B5.1654336
12. Yao C, Webster TJ (2007) Anodization: a promising nano-modification technique for titanium for orthopedic applications. In: Nanotechnology for the regeneration of hard and soft tissues. World Scientific, pp 79–110
13. Yang B, Uchida M, Kim H-M et al (2004) Preparation of bioactive titanium metal via anodic oxidation treatment. Biomaterials 25:1003–1010. https://doi.org/10.1016/S0142-9612(03)00626-4
14. Li G, Ma F, Liu P et al (2023) Review of micro-arc oxidation of titanium alloys: mechanism, properties and applications. J Alloys Compd 948:169773. https://doi.org/10.1016/j.jallcom.2023.169773
15. Zhao J, Wang X, Chen R, Li L (2005) Fabrication of titanium oxide nanotube arrays by anodic oxidation. Solid State Commun 134:705–710. https://doi.org/10.1016/j.ssc.2005.02.028
16. Diamanti MV, Del Curto B, Pedeferri M (2011) Anodic oxidation of titanium: from technical aspects to biomedical applications. J Appl Biomater Biomech 9:55–69. https://doi.org/10.5301/JABB.2011.7429
17. Sul Y-T, Johansson CB, Jeong Y, Albrektsson T (2001) The electrochemical oxide growth behaviour on titanium in acid and alkaline electrolytes. Med Eng Phys 23:329–346. https://doi.org/10.1016/S1350-4533(01)00050-9
18. Choi J, Wehrspohn RB, Lee J, Gösele U (2004) Anodization of nanoimprinted titanium: a comparison with formation of porous alumina. Electrochim Acta 49:2645–2652. https://doi.org/10.1016/j.electacta.2004.02.015

19. Zwilling V, Darque-Ceretti E, Boutry-Forveille A et al (1999) Structure and physicochemistry of anodic oxide films on titanium and TA6V alloy. Surf Interface Anal 27:629–637. https://doi.org/10.1002/(SICI)1096-9918(199907)27:7%3c629::AID-SIA551%3e3.0.CO;2-0

20. Gong D, Grimes CA, Varghese OK et al (2001) Titanium oxide nanotube arrays prepared by anodic oxidation. J Mater Res 16:3331–3334. https://doi.org/10.1557/JMR.2001.0457

21. Beranek R, Hildebrand H, Schmuki P (2003) Self-organized porous titanium oxide prepared in H[sub 2]SO[sub 4]/HF electrolytes. Electrochem Solid-State Lett 6:B12. https://doi.org/10.1149/1.1545192

22. Tsuchiya H, Macak JM, Taveira L et al (2005) Self-organized TiO_2 nanotubes prepared in ammonium fluoride containing acetic acid electrolytes. Electrochem commun 7:576–580. https://doi.org/10.1016/j.elecom.2005.04.008

23. Cai Q, Paulose M, Varghese OK, Grimes CA (2005) The effect of electrolyte composition on the fabrication of self-organized titanium oxide nanotube arrays by anodic oxidation. J Mater Res 20:230–236. https://doi.org/10.1557/JMR.2005.0020

24. Ruan C, Paulose M, Varghese OK et al (2005) Fabrication of highly ordered TiO_2 nanotube arrays using an organic electrolyte. J Phys Chem B 109:15754–15759. https://doi.org/10.1021/jp052736u

25. Mor GK, Varghese OK, Paulose M et al (2003) Fabrication of tapered, conical-shaped titania nanotubes. J Mater Res 18:2588–2593. https://doi.org/10.1557/JMR.2003.0362

26. Bauer S, Park J, von der Mark K, Schmuki P (2008) Improved attachment of mesenchymal stem cells on super-hydrophobic TiO_2 nanotubes. Acta Biomater 4:1576–1582. https://doi.org/10.1016/j.actbio.2008.04.004

27. Le Guéhennec L, Soueidan A, Layrolle P, Amouriq Y (2007) Surface treatments of titanium dental implants for rapid osseointegration. Dent Mater 23:844–854. https://doi.org/10.1016/j.dental.2006.06.025

28. Wei K, Kim B-S, Kim I-S (2011) Fabrication and biocompatibility of electrospun silk biocomposites. Membranes (Basel) 1:275–298. https://doi.org/10.3390/membranes1040275

29. Bayram C, Demirbilek M, Yalçın E et al (2014) Osteoblast response on co-modified titanium surfaces via anodization and electrospinning. Appl Surf Sci 288:143–148. https://doi.org/10.1016/j.apsusc.2013.09.168

30. Liu Q, Ding J, Mante FK et al (2002) The role of surface functional groups in calcium phosphate nucleation on titanium foil: a self-assembled monolayer technique. Biomaterials 23:3103–3111. https://doi.org/10.1016/S0142-9612(02)00050-9

31. Nanci A, Wuest JD, Peru L et al (1998) Chemical modification of titanium surfaces for covalent attachment of biological molecules. J Biomed Mater Res 40:324–335. https://doi.org/10.1002/(SICI)1097-4636(199805)40:2%3c324::AID-JBM18%3e3.0.CO;2-L

32. Cai K, Rechtenbach A, Hao J et al (2005) Polysaccharide-protein surface modification of titanium via a layer-by-layer technique: characterization and cell behaviour aspects. Biomaterials 26:5960–5971. https://doi.org/10.1016/j.biomaterials.2005.03.020

33. Cai K, Hu Y, Jandt KD (2007) Surface engineering of titanium thin films with silk fibroin via layer-by-layer technique and its effects on osteoblast growth behavior. J Biomed Mater Res A 82A:927–935. https://doi.org/10.1002/jbm.a.31233

34. Cai K, Hu Y, Jandt KD, Wang Y (2008) Surface modification of titanium thin film with chitosan via electrostatic self-assembly technique and its influence on osteoblast growth behavior. J Mater Sci Mater Med 19:499–506. https://doi.org/10.1007/s10856-007-3184-5

35. Zhang Z, Jiang T, Ma K et al (2011) Low temperature electrophoretic deposition of porous chitosan/silk fibroin composite coating for titanium biofunctionalization. J Mater Chem 21:7705. https://doi.org/10.1039/c0jm04164e

36. Ravichandran R, Ng CC, Liao S et al (2012) Biomimetic surface modification of titanium surfaces for early cell capture by advanced electrospinning. Biomed Mater 7:015001. https://doi.org/10.1088/1748-6041/7/1/015001

37. Miroiu FM, Socol G, Visan A et al (2010) Composite biocompatible hydroxyapatite–silk fibroin coatings for medical implants obtained by Matrix Assisted Pulsed Laser Evaporation. Mater Sci Eng, B 169:151–158. https://doi.org/10.1016/j.mseb.2009.10.004

38. Shin K, Acri T, Geary S, Salem AK (2017) Biomimetic mineralization of biomaterials using simulated body fluids for bone tissue engineering and regenerative medicine <sup/>. Tissue Eng Part A 23:1169–1180. https://doi.org/10.1089/ten.tea.2016.0556
39. de Jonge LT, Leeuwenburgh SCG, Wolke JGC, Jansen JA (2008) Organic-inorganic surface modifications for titanium implant surfaces. Pharm Res 25:2357–2369. https://doi.org/10.1007/s11095-008-9617-0
40. Kuhn LT (2012) Biomaterials. In: Introduction to biomedical engineering. Elsevier, pp 219–271
41. Bachmann M, Kukkurainen S, Hytönen VP, Wehrle-Haller B (2019) Cell adhesion by integrins. Physiol Rev 99:1655–1699. https://doi.org/10.1152/physrev.00036.2018
42. El-Ghannam A, Ducheyne P, Shapiro IM (1999) Effect of serum proteins on osteoblast adhesion to surface-modified bioactive glass and hydroxyapatite. J Orthop Res 17:340–345. https://doi.org/10.1002/jor.1100170307
43. Ruoslahti E (1996) RGD and other recognition sequences for integrins. Annu Rev Cell Dev Biol 12:697–715. https://doi.org/10.1146/annurev.cellbio.12.1.697
44. Ungai-Salánki R, Peter B, Gerecsei T et al (2019) A practical review on the measurement tools for cellular adhesion force. Adv Colloid Interface Sci 269:309–333. https://doi.org/10.1016/j.cis.2019.05.005
45. Ferrari M, Cirisano F, Morán MC (2019) Mammalian cell behavior on hydrophobic substrates: influence of surface properties. Colloids Interfaces 3:48. https://doi.org/10.3390/colloids3020048
46. Metwally S, Stachewicz U (2019) Surface potential and charges impact on cell responses on biomaterials interfaces for medical applications. Mater Sci Eng C 104:109883. https://doi.org/10.1016/j.msec.2019.109883
47. Liu Y, Rath B, Tingart M, Eschweiler J (2020) Role of implants surface modification in osseointegration: a systematic review. J Biomed Mater Res A 108:470–484. https://doi.org/10.1002/jbm.a.36829
48. Jia L, Han F, Wang H et al (2019) Polydopamine-assisted surface modification for orthopaedic implants. J Orthop Translat 17:82–95. https://doi.org/10.1016/j.jot.2019.04.001
49. Stewart C, Akhavan B, Wise SG, Bilek MMM (2019) A review of biomimetic surface functionalization for bone-integrating orthopedic implants: mechanisms, current approaches, and future directions. Prog Mater Sci 106:100588. https://doi.org/10.1016/j.pmatsci.2019.100588
50. Hu C, Ashok D, Nisbet DR, Gautam V (2019) Bioinspired surface modification of orthopedic implants for bone tissue engineering. Biomaterials 219:119366. https://doi.org/10.1016/j.biomaterials.2019.119366
51. Webster T (2000) Enhanced functions of osteoblasts on nanophase ceramics. Biomaterials 21:1803–1810. https://doi.org/10.1016/S0142-9612(00)00075-2
52. Webster TJ, Ergun C, Doremus RH et al (2000) Specific proteins mediate enhanced osteoblast adhesion on nanophase ceramics. J Biomed Mater Res 51:475–483. https://doi.org/10.1002/1097-4636(20000905)51:3%3c475::AID-JBM23%3e3.0.CO;2-9
53. Rivera-Chacon DM, Alvarado-Velez M, Acevedo-Morantes CY et al (2013) Fibronectin and vitronectin promote human fetal osteoblast cell attachment and proliferation on nanoporous titanium surfaces. J Biomed Nanotechnol 9:1092–1097. https://doi.org/10.1166/jbn.2013.1601
54. Khang D, Kim SY, Liu-Snyder P et al (2007) Enhanced fibronectin adsorption on carbon nanotube/poly(carbonate) urethane: independent role of surface nano-roughness and associated surface energy. Biomaterials 28:4756–4768. https://doi.org/10.1016/j.biomaterials.2007.07.018
55. Woo KM, Chen VJ, Ma PX (2003) Nano-fibrous scaffolding architecture selectively enhances protein adsorption contributing to cell attachment. J Biomed Mater Res A 67A:531–537. https://doi.org/10.1002/jbm.a.10098
56. Bowers DT, Brown JL (2019) Nanofibers as bioinstructive scaffolds capable of modulating differentiation through mechanosensitive pathways for regenerative engineering. Regen Eng Transl Med 5:22–29. https://doi.org/10.1007/s40883-018-0076-9

57. Anselme K, Bigerelle M, Noël B et al (2002) Effect of grooved titanium substratum on human osteoblastic cell growth. J Biomed Mater Res 60:529–540. https://doi.org/10.1002/jbm.10101
58. Wall I, Donos N, Carlqvist K et al (2009) Modified titanium surfaces promote accelerated osteogenic differentiation of mesenchymal stromal cells in vitro. Bone 45:17–26. https://doi.org/10.1016/j.bone.2009.03.662
59. Zanicotti D, Duncan W, Seymour G, Coates D (2018) Effect of titanium surfaces on the osteogenic differentiation of human adipose-derived stem cells. Int J Oral Maxillofac Implants 33:e77–e87. https://doi.org/10.11607/jomi.5810
60. Fujibayashi S, Neo M, Kim H-M et al (2004) Osteoinduction of porous bioactive titanium metal. Biomaterials 25:443–450. https://doi.org/10.1016/S0142-9612(03)00551-9
61. Kim M, Kim C, Lim Y, Heo S (2006) Microrough titanium surface affects biologic response in MG63 osteoblast-like cells. J Biomed Mater Res A 79A:1023–1032. https://doi.org/10.1002/jbm.a.31040
62. Peng L, Wu F, Cao M et al (2023) Effects of different physical factors on osteogenic differentiation. Biochimie 207:62–74. https://doi.org/10.1016/j.biochi.2022.10.020
63. Salazar GT, Ohneda O (2013) Review of biophysical factors affecting osteogenic differentiation of human adult adipose-derived stem cells. Biophys Rev 5:11–28. https://doi.org/10.1007/s12551-012-0079-6
64. Lv H, Li L, Sun M et al (2015) Mechanism of regulation of stem cell differentiation by matrix stiffness. Stem Cell Res Ther 6. https://doi.org/10.1186/s13287-015-0083-4
65. Griffin MF (2015) Control of stem cell fate by engineering their micro and nanoenvironment. World J Stem Cells 7:37. https://doi.org/10.4252/wjsc.v7.i1.37
66. Rahmati M, Silva EA, Reseland JE et al (2020) Biological responses to physicochemical properties of biomaterial surface. Chem Soc Rev 49:5178–5224. https://doi.org/10.1039/D0CS00103A
67. Kolind K, Leong KW, Besenbacher F, Foss M (2012) Guidance of stem cell fate on 2D patterned surfaces. Biomaterials 33:6626–6633. https://doi.org/10.1016/j.biomaterials.2012.05.070
68. Dalby MJ, Gadegaard N, Oreffo ROC (2014) Harnessing nanotopography and integrin–matrix interactions to influence stem cell fate. Nat Mater 13:558–569. https://doi.org/10.1038/nmat3980
69. Chen C-S, Chang J-H, Srimaneepong V et al (2020) Improving the in vitro cell differentiation and in vivo osseointegration of titanium dental implant through oxygen plasma immersion ion implantation treatment. Surf Coat Technol 399:126125. https://doi.org/10.1016/j.surfcoat.2020.126125
70. Le PTM, Shintani SA, Takadama H et al (2021) Bioactivation treatment with mixed acid and heat on titanium implants fabricated by selective laser melting enhances preosteoblast cell differentiation. Nanomaterials 11:987. https://doi.org/10.3390/nano11040987
71. Li K, Liu S, Hu T et al (2020) Optimized nanointerface engineering of micro/nanostructured titanium implants to enhance cell-nanotopography interactions and osseointegration. ACS Biomater Sci Eng 6:969–983. https://doi.org/10.1021/acsbiomaterials.9b01717
72. Wu B, Tang Y, Wang K et al (2022) Nanostructured titanium implant surface facilitating osseointegration from protein adsorption to osteogenesis: the example of TiO_2 NTAs. Int J Nanomed 17:1865–1879. https://doi.org/10.2147/IJN.S362720
73. Wang Y, Tang S, Ding N, Zhang Z (2023) Biological properties of hydroxyapatite coatings on titanium dioxide nanotube surfaces using negative pressure method. J Biomed Mater Res B Appl Biomater 111:1365–1373. https://doi.org/10.1002/jbm.b.35240
74. Esteban J, Cordero-Ampuero J (2011) Treatment of prosthetic osteoarticular infections. Expert Opin Pharmacother 12:899–912. https://doi.org/10.1517/14656566.2011.543676
75. Miclau T, Schmidt AH, Wenke JC et al (2010) Infection. J Orthop Trauma 24:583–586. https://doi.org/10.1097/BOT.0b013e3181eebf12
76. Antoci V, Adams CS, Parvizi J et al (2008) The inhibition of Staphylococcus epidermidis biofilm formation by vancomycin-modified titanium alloy and implications for the treatment of periprosthetic infection. Biomaterials 29:4684–4690. https://doi.org/10.1016/j.biomaterials.2008.08.016

77. Antoci V, King SB, Jose B et al (2007) Vancomycin covalently bonded to titanium alloy prevents bacterial colonization. J Orthop Res 25:858–866. https://doi.org/10.1002/jor.20348
78. Sia IG, Berbari EF, Karchmer AW (2005) Prosthetic joint infections. Infect Dis Clin North Am 19:885–914. https://doi.org/10.1016/j.idc.2005.07.010
79. Zilberman M, Elsner J (2008) Antibiotic-eluting medical devices for various applications. J Control Release 130:202–215. https://doi.org/10.1016/j.jconrel.2008.05.020
80. Lentino JR (2003) Prosthetic joint infections: bane of orthopedists, challenge for infectious disease specialists. Clin Infect Dis 36:1157–1161. https://doi.org/10.1086/374554
81. Campoccia D, Montanaro L, Arciola CR (2006) The significance of infection related to orthopedic devices and issues of antibiotic resistance. Biomaterials 27:2331–2339. https://doi.org/10.1016/j.biomaterials.2005.11.044
82. Zhao L, Chu PK, Zhang Y, Wu Z (2009) Antibacterial coatings on titanium implants. J Biomed Mater Res B Appl Biomater 91B:470–480. https://doi.org/10.1002/jbm.b.31463
83. Pesode PA, Barve SB (2021) Recent advances on the antibacterial coating on titanium implant by micro-Arc oxidation process. Mater Today Proc 47:5652–5662. https://doi.org/10.1016/j.matpr.2021.03.702
84. Akshaya S, Rowlo PK, Dukle A, Nathanael AJ (2022) Antibacterial coatings for titanium implants: recent trends and future perspectives. Antibiotics 11:1719. https://doi.org/10.3390/antibiotics11121719
85. Jaggessar A, Senevirathne SWMAI, Velic A, Yarlagadda PKDV (2022) Antibacterial activity of 3D versus 2D TiO_2 nanostructured surfaces to investigate curvature and orientation effects. Curr Opin Biomed Eng 23:100404. https://doi.org/10.1016/j.cobme.2022.100404
86. Prakash J, Cho J, Mishra YK (2022) Photocatalytic TiO_2 nanomaterials as potential antimicrobial and antiviral agents: scope against blocking the SARS-COV-2 spread. Micro Nano Eng 14:100100. https://doi.org/10.1016/j.mne.2021.100100
87. Yang X, Hou J, Tian Y et al (2022) Antibacterial surfaces: strategies and applications. Sci China Technol Sci 65:1000–1010. https://doi.org/10.1007/s11431-021-1962-x
88. Hasan J, Crawford RJ, Ivanova EP (2013) Antibacterial surfaces: the quest for a new generation of biomaterials. Trends Biotechnol 31:295–304. https://doi.org/10.1016/j.tibtech.2013.01.017
89. Bright R, Fernandes D, Wood J et al (2022) Long-term antibacterial properties of a nanostructured titanium alloy surface: an in vitro study. Mater Today Bio 13:100176. https://doi.org/10.1016/j.mtbio.2021.100176
90. Nikpasand A, Parvizi MR (2019) Evaluation of the effect of titatnium dioxide nanoparticles/gelatin composite on infected skin wound healing; an animal model study. Bull Emerg Trauma 7:366–372. https://doi.org/10.29252/beat-070405
91. Cao Y, Su B, Chinnaraj S et al (2018) Nanostructured titanium surfaces exhibit recalcitrance towards staphylococcus epidermidis biofilm formation. Sci Rep 8:1071. https://doi.org/10.1038/s41598-018-19484-x
92. Hajkova P, Spatenka P, Horsky J et al (2007) Photocatalytic effect of TiO_2 films on viruses and bacteria. Plasma Processes Polym 4:S397–S401. https://doi.org/10.1002/ppap.200731007
93. Nam Y, Lim JH, Ko KC, Lee JY (2019) Photocatalytic activity of TiO_2 nanoparticles: a theoretical aspect. J Mater Chem A Mater 7:13833–13859. https://doi.org/10.1039/C9TA03385H
94. Reghunath S, Pinheiro D, KR SD (2021) A review of hierarchical nanostructures of TiO_2: advances and applications. Appl Surf Sci Adv 3:100063. https://doi.org/10.1016/j.apsadv.2021.100063
95. Linley S, Thomson NR (2021) Environmental applications of nanotechnology: nano-enabled remediation processes in water, soil and air treatment. Water Air Soil Pollut 232:59. https://doi.org/10.1007/s11270-021-04985-9
96. He D, Zhang X, Yao X, Yang Y (2022) In vitro and in vivo highly effective antibacterial activity of carbon dots-modified TiO_2 nanorod arrays on titanium. Colloids Surf B Biointerfaces 211:112318. https://doi.org/10.1016/j.colsurfb.2022.112318
97. Li Z, Wang E, Zhang Y, Luo R, Gai Y, Ouyang H, Feng H et al (2023) Antibacterial ability of black titania in dark: via oxygen vacancies mediated electron transfer. Nano Today 50:101826. https://doi.org/10.1016/j.nantod.2023.101826

98. Wang Q, Huang J-Y, Li H-Q et al (2016) Recent advances on smart TiO_2 nanotube platforms for sustainable drug delivery applications. Int J Nanomed 12:151–165. https://doi.org/10. 2147/IJN.S117498

99. Lai Y-K, Wang Q, Huang J-Y et al (2016) TiO_2 nanotube platforms for smart drug delivery: a review. Int J Nanomed 11:4819–4834. https://doi.org/10.2147/IJN.S108847

100. Jafari S, Mahyad B, Hashemzadeh H et al (2020) Biomedical applications of TiO_2 nanostructures: recent advances. Int J Nanomed 15:3447–3470. https://doi.org/10.2147/IJN.S24 9441

101. Fathi M, Akbari B, Taheriazam A (2019) Antibiotics drug release controlling and osteoblast adhesion from Titania nanotubes arrays using silk fibroin coating. Mater Sci Eng C 103:109743. https://doi.org/10.1016/j.msec.2019.109743

102. Keceli HG, Bayram C, Celik E et al (2020) Dual delivery of platelet-derived growth factor and bone morphogenetic factor-6 on titanium surface to enhance the early period of implant osseointegration. J Periodontal Res 55:694–704. https://doi.org/10.1111/jre.12756

103. Popat KC, Eltgroth M, LaTempa TJ et al (2007) Decreased Staphylococcus epidermis adhesion and increased osteoblast functionality on antibiotic-loaded titania nanotubes. Biomaterials 28:4880–4888. https://doi.org/10.1016/j.biomaterials.2007.07.037

104. Yao C, Webster TJ (2009) Prolonged antibiotic delivery from anodized nanotubular titanium using a co-precipitation drug loading method. J Biomed Mater Res B Appl Biomater 91B:587–595. https://doi.org/10.1002/jbm.b.31433

105. Ma M, Kazemzadeh-Narbat M, Hui Y et al (2012) Local delivery of antimicrobial peptides using self-organized TiO_2 nanotube arrays for peri-implant infections. J Biomed Mater Res A 100A:278–285. https://doi.org/10.1002/jbm.a.33251

106. Gulati K, Ramakrishnan S, Aw MS et al (2012) Biocompatible polymer coating of titania nanotube arrays for improved drug elution and osteoblast adhesion. Acta Biomater 8:449–456. https://doi.org/10.1016/j.actbio.2011.09.004

107. Peng L, Mendelsohn AD, LaTempa TJ et al (2009) Long-term small molecule and protein elution from TiO_2 nanotubes. Nano Lett 9:1932–1936. https://doi.org/10.1021/nl9001052

108. Bayram C (2022) Prolonged biomolecule release from titanium surfaces via titania nanotube arrays. Celal Bayar Üniversitesi Fen Bilimleri Dergisi 18:1–7. https://doi.org/10.18466/cba yarfbe.972316

109. Çalışkan N, Bayram C, Erdal E et al (2014) Titania nanotubes with adjustable dimensions for drug reservoir sites and enhanced cell adhesion. Mater Sci Eng C 35:100–105. https://doi. org/10.1016/j.msec.2013.10.033

110. Sinn Aw M, Kurian M, Losic D (2014) Non-eroding drug-releasing implants with ordered nanoporous and nanotubular structures: concepts for controlling drug release. Biomater Sci 2:10–34. https://doi.org/10.1039/C3BM60196J

111. Simovic S, Losic D, Vasilev K (2010) Controlled drug release from porous materials by plasma polymer deposition. Chem Commun 46:1317. https://doi.org/10.1039/b919840g

112. Simovic S, Diener KR, Bachhuka A et al (2014) Controlled release and bioactivity of the monoclonal antibody rituximab from a porous matrix: a potential in situ therapeutic device. Mater Lett 130:210–214. https://doi.org/10.1016/j.matlet.2014.05.110

113. Aw MS, Gulati K, Losic D (2011) Controlling drug release from titania nanotube arrays using polymer nanocarriers and biopolymer coating. J Biomater Nanobiotechnol 02:477–484. https://doi.org/10.4236/jbnb.2011.225058

114. Aw MS, Addai-Mensah J, Losic D (2012) Polymer micelles for delayed release of therapeutics from drug-releasing surfaces with nanotubular structures. Macromol Biosci 12:1048–1052. https://doi.org/10.1002/mabi.201200012

115. Cai K, Jiang F, Luo Z, Chen X (2010) Temperature-responsive controlled drug delivery system based on titanium nanotubes. Adv Eng Mater 12. https://doi.org/10.1002/adem.201080015

116. Shrestha NK, Macak JM, Schmidt-Stein F et al (2009) Magnetically guided titania nanotubes for site-selective photocatalysis and drug release. Angew Chem Int Ed 48:969–972. https:// doi.org/10.1002/anie.200804429

117. Aw MS, Addai-Mensah J, Losic D (2012) Magnetic-responsive delivery of drug-carriers using Titania nanotube arrays. J Mater Chem 22:6561. https://doi.org/10.1039/c2jm16819g
118. Ainslie KM, Tao SL, Popat KC et al (2009) In vitro inflammatory response of nanostructured Titania, silicon oxide, and polycaprolactone. J Biomed Mater Res A 91A:647–655. https://doi.org/10.1002/jbm.a.32262
119. Losic D, Velleman L, Kant K et al (2011) Self-ordering electrochemistry: a simple approach for engineering nanopore and nanotube arrays for emerging applications. Aust J Chem 64:294. https://doi.org/10.1071/CH10398

Smart Piezoelectric Materials for Hard and Cartilage Tissue Repair and Reconstruction

Sevin Adiguzel, Serap Sezen, and Feray Bakan Misirlioglu

Abstract Piezoelectricity is a property unique to certain classes of materials where mechanical strain can be converted to electrical potential and vice versa. The unique capabilities of piezoelectric materials offer significant potential in tissue engineering, specifically in bone and cartilage regeneration. That piezoelectric materials can convert mechanical input into electrical signals which provides a highly convenient and beneficial path to facilitate cell growth by influencing the existing signaling channels. The versatility of piezoelectric materials in creating scaffolds for tissue development and sensors for monitoring tissue regeneration is also invaluable. This chapter comprehensively explores the utilization of various piezoelectric materials, some important forth coming studies in both inorganic and organic materials, in the regeneration of bone and cartilage tissues.

Keywords Piezoelectric · Smart materials · Tissue engineering · Tissue regeneration · Bone · Cartilage

1 Introduction

Piezoelectric materials exhibit the capacity to produce an electric charge in response to external mechanical strain or, conversely, to transform mechanical stimulation into bioelectrical signals. This inherent property can be purposefully tailored to facilitate new cell growth and proliferation by influencing existing electrical signaling pathways. Such an adaptability makes these materials invaluable in tissue engineering (TE), particularly bone and cartilage. They play a pivotal role in creating scaffolds that provide electrical stimuli to promote tissue development and in crafting sensors that monitor the progress of tissue regeneration. The utilization of piezoelectric materials

S. Adiguzel · S. Sezen
Faculty of Engineering and Natural Sciences, Sabanci University, Istanbul, Turkey

F. B. Misirlioglu (✉)
Nanotechnology Research and Application Center (SUNUM), Sabanci University, Istanbul, Turkey
e-mail: feraybakan@sabanciuniv.edu

© The Author(s), under exclusive license to Springer Nature Singapore Pte Ltd. 2025   59
N. Sağlam et al. (eds.), *Nano-Biomaterials in Tissue Repair and Regeneration*, Tissue Repair and Reconstruction, https://doi.org/10.1007/978-981-96-1341-0_3

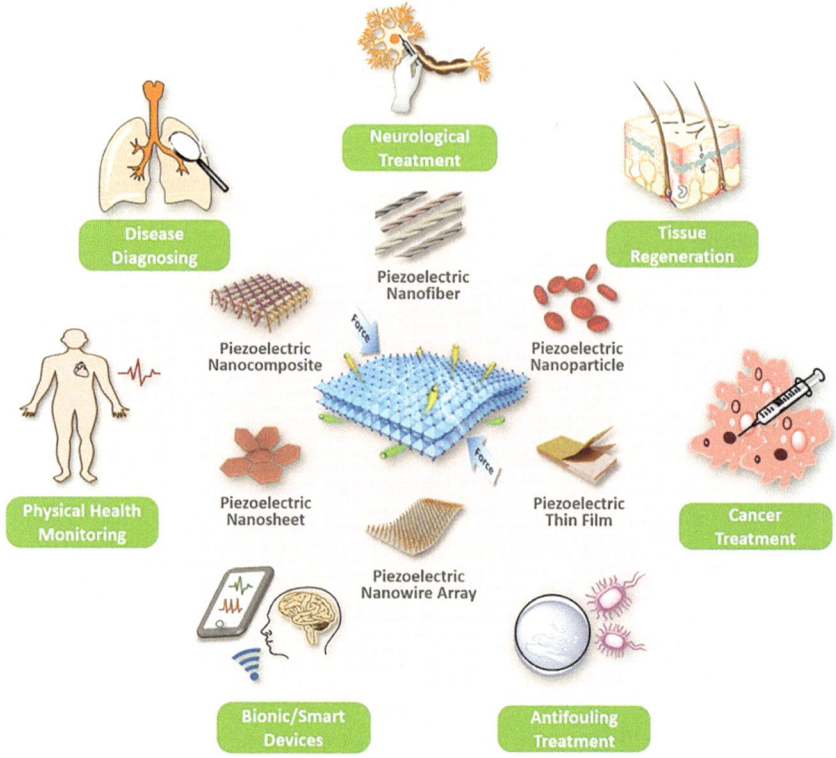

Fig. 1 Demonstration of the use of different piezoelectric platforms in biomedical applications. Adapted with the permission of Ref. [1]. Copyright, 2021, John Wiley & Sons

in TE confers numerous advantages over conventional materials, including biocompatibility, robustness, versatility, electrical stimulation capabilities, and monitoring functionality. Their defining attribute lies in their ability to generate electrical charge when subjected to mechanical stress or deformation. This unique trait underpins tissue regeneration in both engineering and biomedical realms without necessitating an external power source. For a comprehensive overview of the diverse biomedical applications of various forms of piezoelectric materials, please refer to Fig. 1 [1].

2 Definition of Piezoelectric Effect

The term piezoelectricity originates from the Greek word "piezin", signifying "to press" and is a property of certain materials that can develop an electrical voltage under mechanical load [2]. Crystalline materials with non-symmetrical arrangement

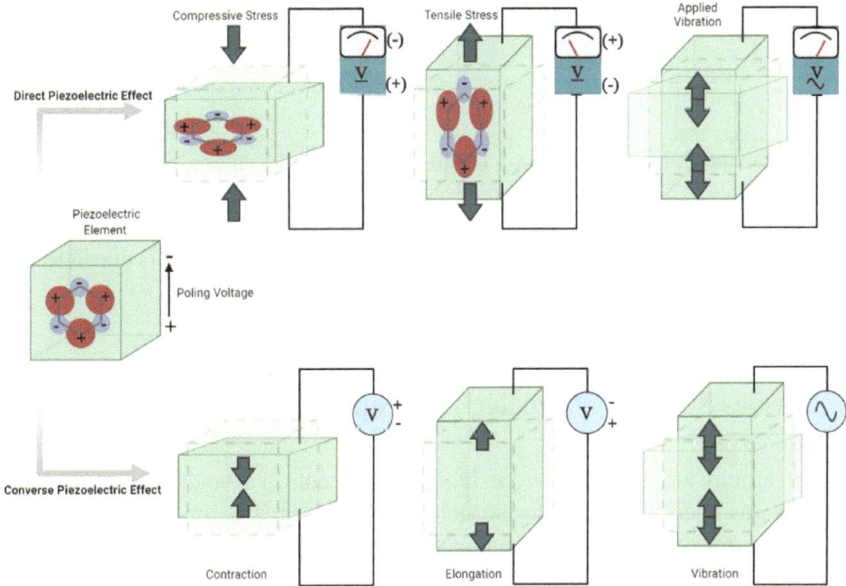

Fig. 2 Schematic representation of direct and converse piezoelectric effect. Adapted with the permission of Ref. [5]. Copyright, 2021, Mary Ann Liebert, Inc.

of ions resulting in polar structures lack a center of symmetry and display piezo-electric characteristics. These materials constitute 20 out of the 32 crystallographic point groups, with half of them exhibiting induced electrical polarization in response to mechanical stress or strain. The remaining ten groups possess spontaneous polar-ization and can show piezoelectric, pyroelectric, and ferroelectric effects, where piezoelectric crystals may exhibit pyroelectric and ferroelectric properties as well [3]. One should note that all ferroelectric crystals are piezoelectric and pyroelectrics, however the reverse is not necessarily applicable.

In 1880, French physicists Jacques and Pierre Curie were the first to discover direct piezoelectricity in single-crystal quartz. The immediate effect illustrates the ability of specific materials to produce electric charges with opposite polarities in between crystal surfaces when subjected to an external force that induces mechanical deformation [4]. Subsequently, in 1881, Gabriel Lippmann observed the converse piezoelectric effect. The converse piezoelectric effect is related to the structural alterations in the crystal when exposed to an external electric field. The realignment of dipolar moments in response to the electric field results in the deformation of the crystal, and this deformation is directly proportional to the strength of the applied electric field via what one calls the converse piezoelectric tensor of the crystal. Both effects, namely, the direct and converse piezoelectric, are explained by the tensor relations (Eqs. 1 and 2) shown below, and the schematic representation was shown in Fig. 2. These equations connect the electrical displacement to the stress and electric field applied to the material.

$$D_i = d_{ikl}T_{kl} + \varepsilon_{ik}^T E^k \tag{1}$$

$$S_{ij} = s_{ijkl}^E T_{kl} + d_{kij}E_k \tag{2}$$

In these equations, subscripts $i, j, k,$ and l take 1, 2, and 3 values. Electric displacement and electric field are represented by D and E, and mechanical strain and stress are represented by S and T, respectively. Additionally, d is the direct piezoelectric tensor, s^E is the elastic compliance in a constant electric field, and ε^T is the dielectric constant at a given stress [3].

3   Tissues Showing Piezoelectric Properties

Piezoelectricity was first observed in biological systems in a bundle of wool in 1940 [6], and was later coined as bio-piezoelectricity, which is an important discovery in living systems. Bio-piezoelectricity is vital in maintaining many critical cellular functions, including cell division, proliferation, migration, differentiation, intracellular communication, protein folding, and chemotaxis. Moreover, it affects neuronal activities, ion channels, heart and muscle contraction as well as bone and epithelial healing [7]. Piezoelectricity can be observed in many tissues, including hard tissues such as bone, dentin, and cementum, and soft tissues such as cartilage, ligament, tendon, muscle, hair, skin, and pineal gland in the human body. The piezoelectric origin of these tissues is likely related to the inherent polarization arising from the low symmetry; lack of inversion center; and highly ordered crystal structure of extracellular matrix (ECM) components such as hydroxyapatite, glycosaminoglycans (GAGs), collagen, elastin, and keratin [8, 9].

As revealed by Fukada in 1957, bone can be considered as a composite that shows direct and converse piezoelectric effects, as it is packed in the form of non-centrosymmetric collagen matrix (22 wt.%), fibrils and contains hydroxyapatite (HA) crystals (69 wt.%) embedded in it [10, 11]. The piezoelectric property of bone originates from collagen type I in its structure, and the piezoelectric coefficient is close to 0.7 pC/N (picocoulomb per Newton). It has a triple helix structure that connects three polypeptide strands by hydrogen bonding, acting as if it has a crystal structure and provides a piezoelectric effect by directing dipoles with −NH and −CO groups [12]. When a mechanical force is applied to bone tissue, collagen fibers slide over each other, causing dipole rearrangement resulting in the separation and polarization of charged groups [13]. This property causes piezoelectric stimulations in bone cells such as osteoblasts and osteocytes. Such a process greatly impacts the formation of an endogenous electric field that affects osteogenesis. As shown in Fig. 3, with tissue compression, the surface charge density in the bone matrix increases, and a large zeta potential occurs. Thus, the flow potential increases, contributing to the endogenous electric field [13].

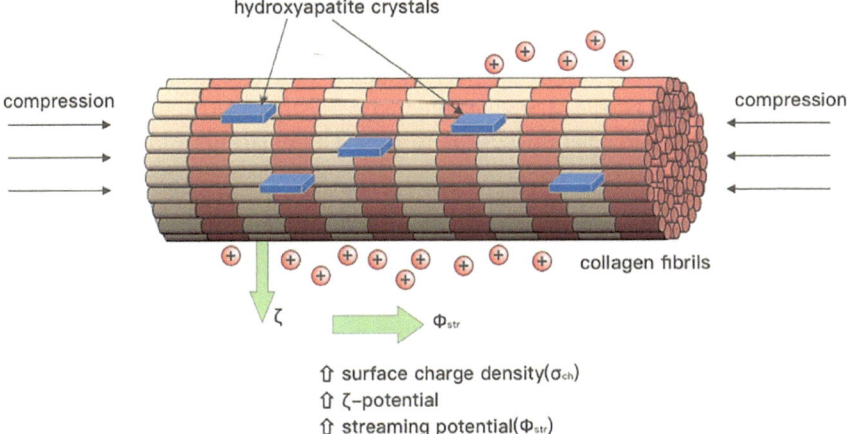

hydroxyapatite crystals

compression

compression

collagen fibrils

ζ Φ_{str}

⇧ surface charge density(σ_{ch})
⇧ ζ–potential
⇧ streaming potential(Φ_{str})

Fig. 3 Schematic representation of the mechanism of formation of electrical effects in bone tissue due to flow potential. Adapted with the permission of Ref. [13]. Copyright, 2023, John Wiley & Sons

Cartilage tissue also has piezoelectric properties due to collagen, which creates streaming electric potential under mechanical stress. Similar to bone, cartilage is dense connective tissue comprising cells and ECM. It has a more flexible ECM than bone due to its different cells and different collagen types (II, VI, IX, X, and XI), proteoglycan, non-collagenous proteins, and tissue fluid. Moreover, tendon and ligament tissues show piezoelectric properties. The tendon is dense, connective, and load-bearing tissue that connects muscle to bone, and the ligaments connect bone to bone and sustain joint stability [14]. Like bone and cartilage, tendon and ligament tissues have highly hierarchical structures and piezoelectric collagen fibrils responsible for mechanotransduction by aligning in the direction of mechanical loading. When these tissues are exposed to mechanical stress, mechanotransduction mechanisms that enable the development and healing of tissues are activated [15, 16]. The cornea and sclera display piezoelectric features because of the collagen inherent. In the cornea, collagen is orthogonally aligned, but it is randomly oriented in the sclera. Due to this difference, the piezoelectric response of the sclera is lower than that of the cornea. Also, piezoelectricity varies with hydration level and reduces with dehydration [11]. The skin also possesses a piezoelectric behavior: a quantitative study revealed that the piezoelectric constant of the dermis layer was found to be 0.1 C/N, and originated from the piezoelectric feature of collagen. For other layers, the piezoelectric effect caused by the keratin and piezoelectric constant of the epidermis and horny layer were found to be 0.03 pC/N and 0.2 pC/N, respectively [17].

The electrical attributes present in these tissues, identified as piezoelectric tissue owing to the presence of piezoelectric proteins within the ECM, significantly influence tissue functionality. Consequently, the piezoelectric quality operates akin to bioelectricity, facilitating the generation of electrical signals through proteins like collagen, which, in turn, trigger crucial intracellular signaling cascades essential for cellular functions such as growth, matrix synthesis, and tissue restoration [18].

The mechanisms by which electrical stimulation affects cell function in these tissues are not fully known, although some processes have already been identified (Fig. 4). The most obvious of these is to trigger the activation of voltage-gated Ca^{2+} channels by changing the resting membrane potential of the cells and, accordingly, promoting calcium uptake into the cell. Increasing the calcium level within the cell triggers the calcineurin and calmodulin pathways, changing the gene expression profile of the cells and encouraging the production of growth factors (such as TGF-β, BMP-2) [19]. These growth factors play an important role in the formation of bone and cartilage [18, 20]. Another potential mechanism is the activation of mitogen-activated protein kinase (MAPK) signaling pathways that promote actin rearrangement and reorganization. This situation affects the cytoskeleton and increases adenosine triphosphate (ATP) production [19, 21, 22].

As the human population ages, degenerative conditions, particularly bone and cartilage degeneration, have emerged as significant concerns. Tissue engineering (TE) strategies utilizing intelligent materials offer a solution to challenges inherent in conventional treatments—such as drug toxicity, absence of precise targeting, potential complications, and susceptibility to infections [23]. Piezoelectric materials are promising candidates among functional materials that find use for bone- and cartilage-related problems. That these materials can generate electrical stimulation in response to mechanical stress and vice versa is highly beneficial to initiate tissue regeneration at the implantation site without needing external power sources or electrodes [18, 23, 24].

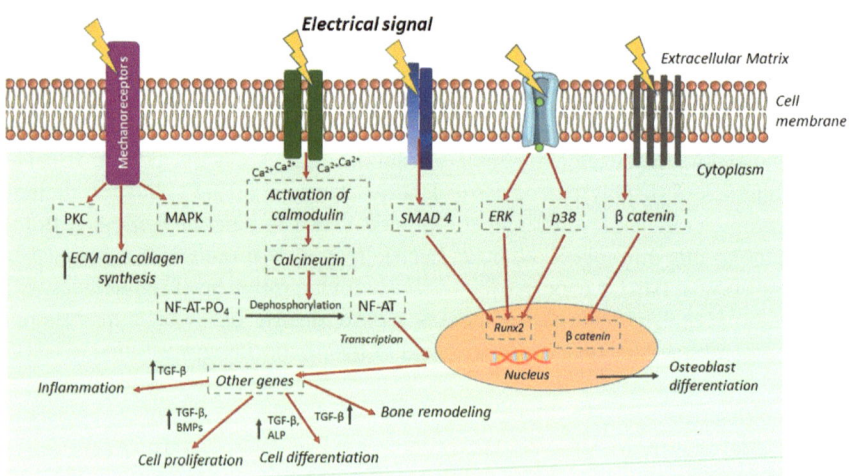

Fig. 4 Schematic representation of activation of signal transduction pathways in response to conversion of mechanical stimulation to electrical stimulation in bone tissue. Redrawn from Ref. [17]

4 Use of Piezoelectric Materials in Bone and Cartilage Regeneration

The converse piezoelectric response of piezoelectric materials can aid in enhancement of tissue growth and healing [25]. Crafting a scaffold demands a nuanced consideration of the mechanical attributes of bone and cartilage and Young's modulus. Precise adjustment of mechanical loading becomes pivotal to avert potential damage. Hence, the material selection necessitates a balance—neither excessively pliant nor excessively rigid. Leveraging the preferred piezoelectric material optimizes the converse piezoelectric effect, furnishing the precise mechanical stimulus to prevent such damage [20, 26]. The efficacy of piezoelectric materials employed as scaffolds heavily relies on their design, morphology, and piezoelectric traits in eliciting the desired cellular responses [24]. In the following sub-sections, various piezoelectric materials, such as spanning polymers, ceramics, and composites, have been scrutinized as bioactive scaffolds for the regeneration of bone and cartilage.

4.1 Piezoelectric Polymers

Synthetic and natural piezoelectric polymers are viable options for bone and cartilage regeneration owing to their inherent piezoelectricity and biocompatibility. Poly (vinylidene fluoride) (PVDF) encompassing β, γ, and δ phases, poly (vinylidene fluoride/trifluoroethylene) (P(VDF/TrFE)), poly (L-lactic acid) (PLLA), polyhydroxybutyrate (PHB), and polyamides (such as nylons and peptides) are among the synthetic variants. On the other hand, natural piezoelectric polymers like polysaccharides (cellulose, amylose, and chitosan) and proteins (such as collagen, silk, and keratin) possess inherent piezoelectric properties. Fabricating scaffolds containing these polymers involves various methodologies, including fiber electrospinning, 3D printing, solvent casting, particulate leaching, compression molding, gel casting, gas foaming, and selective laser sintering [27].

PVDF, renowned for its versatility with five distinct polymorphs, is a prominent piezoelectric polymer in tissue engineering, boasting a piezoelectric coefficient of 20 pC/N. Its widespread adoption in TE applications is shown to stem from its exceptional flexibility, resistance, non-toxic properties, and notably high piezoelectric coefficient [28]. The biocompatibility of PVDF significantly influences cell adhesion and proliferation, correlating with factors like hydrophilicity, hydrophobicity, poling, and scaffold morphology. Notably, the piezoelectric effect inherent in PVDF's polarized surfaces is thought to be playing a pivotal role in cell differentiation and augments bone growth, showcasing its osteogenic potential [18, 27, 29]. Specifically, under dynamic conditions, electroactive poled-PVDF demonstrates heightened osteogenic differentiation in human adipose stem cells [30].

P(VDF/TrFE) emerges as a copolymer derived from vinylidene fluoride (VDF) and trifluoroethylene (TrFE), boasting the highest piezoelectric coefficient, reported to reach values as high as 30 pC/N [31]. When the content ratio of the monomers PVDF and TrFE falls within a suitable range, the random copolymers crystallize into a polar crystal form isomorphic to the β-form of PVDF, and thus exhibits an exceptional piezoelectric behavior among piezoelectric polymers [32]. Beyond its piezoelectric prowess, P(VDF/TrFE) exhibits remarkable biocompatibility, fostering an environment conducive to cell adhesion and proliferation while demonstrating osteogenic and chondrogenic capabilities [28, 33, 34]. When fashioned into a 3D scaffold via electrospinning, P(VDF/TrFE) under mechanical stimulation significantly augments proliferation, differentiation, ECM mineralization, and gene expression in human mesenchymal stem cells (MSC) [34]. Exploring combinations with other piezoelectric materials like BaTiO$_3$, the study on hMSC cells revealed amplified cellular metabolic activity and stimulated neo-bone formation [27]. The electrospun, nanofibrous-based scaffold form of P(VDF/TrFE) copolymer has been shown to be very beneficial in regenerating cartilage tissue effectively [35]. The piezoelectric fibers are preferred to stimulate differentiable cells to mature and to support stem-cell-induced tissue repair [36]. It was reported that the differentiation of hMSCs into chondrocytes was promoted using a piezoelectric scaffold comprising P(VDF/TrFE) fiber [37].

Poly (L-lactic acid) (PLLA) is another synthetic polymer that contains −CO dipoles in its structure that asymmetrically displace under mechanical stress and thus have piezoelectric (d_{14} ~ 10pC/N) properties [27, 28]. PLLA is FDA-approved polymer and, due to its strong mechanical properties, very commonly used in orthopedic devices (screws, pins, and plates) [28]. It can be designed to exhibit slower degradation, thus preventing stress-induced bone atrophy. Moreover, electrospun PLLA nanofibers could be used as a matrix in bone implants and grafts due to their ability to mimic ECM [5]. PLLA scaffold, which mimics bone and contains hydroxapatite and collagen, stimulates the metabolic activities of SaOS-2 and hFOB cells by providing a bone-like environment [38, 39]. It was demonstrated that the nanofiber composite scaffold created by the electrospinning method using PLLA and nano-sized demineralized bone powder (DBP) was able to provide osteogenic differentiation in the hMSC cell line, aiding in skull defect recovery in the rat model [40]. PLLA scaffolds are also used for the stimulation of cartilage-like tissue formation. The biodegradable PLLA can expedite cartilage regeneration owing to its inherent piezoelectric attributes [41].

PHB and its copolymer PHBV exhibit piezoelectricity (d_{14} ~ 1.3 pC/N) attributed to the polar −CO structures, drawing interest due to their biodegradable and biocompatible traits. Nevertheless, owing to its limited mechanical strength and insolubility in water, PHB is typically utilized in tissue engineering by combining it with other polymers, such as PCL [27, 33].

Given the benefits offered by the PHBV copolymer—its biocompatibility, low toxicity, and extended degradation timeline—it is an excellent material for employment in the regeneration of cartilage and bone. The piezoelectric fibrous scaffolds crafted from PHBV, infused with hydroxyapatite (HA), showcased remarkable osteogenic potential, fostering enhanced cell adhesion and differentiation in hMSCs. These composite fibrous scaffolds, exhibiting piezoelectric properties, displayed elevated calcium accumulation, underscoring their promise as a potential structure for bone tissue engineering [42]. Moreover, there's a marked potential for advancing cartilage regeneration by employing the electrostatic spinning technique to stimulate PHBV doped with $BaTiO_3$, a piezoelectric scaffold. This scaffold notably boosted the proliferation, migration, and expression of the collagen II gene in hMSC-derived chondrocytes [43].

Natural piezoelectric polymers have gained great interest in TE due to their low toxicity, biodegradability, and promote cell activities. Cellulose is the most common natural piezoelectric polymer (d_{14} ~ 0.2 pC/N) on Earth and is mostly obtained from bacteria, tunicates, and plants. It is a natural piezoelectric polymer widely used in the medical field (such as tissue repair, cortical, and medical implants) due to its hydrophilicity, nanofiber structure, high tensile strength, cytocompatibility, and bioactivity [41]. Moreover, it is a strong scaffold candidate for bone and cartilage regeneration, besides good biocompatibility and mechanical properties owing to its ability to promote cellular adhesion of chondrocytes and osteoblasts [44, 45]. Microporous bacterial cellulose (BC) scaffolds were fabricated to facilitate the ingrowth of osteoblasts and mineralized tissue formation, and their seeding with MC3T3-E1 osteoprogenitor cells showed positive results regarding cell distribution and mineral accumulation [44]. In another study conducted to examine cartilage regeneration, three-dimensional (3D) porous decellularized cartilage ECM (DCECM) scaffolds that structurally and biochemically mimic the cartilage regeneration environment were produced for the use of bacterial cellulose in cartilage tissue regeneration. BC/DCECM scaffolds exhibited improved cell adhesion and proliferation and in vivo and in vitro studies demonstrated that neocartilage tissue regeneration was achieved [46].

Chitosan, an additional biodegradable and biocompatible piezoelectric polysaccharide (0.2–1.5 pC/N), finds extensive use in biomedical applications owing to its hydrophilic nature, fostering cell adhesion, proliferation, and differentiation [47]. While its mechanical strength might be limited, it is suitable for application in composite form for bone and cartilage regeneration. When aiming at bone tissue regeneration, creating porous hybrid membranes using chitosan and calcium phosphates aids in cell adhesion, proliferation, and differentiation of osteoblast cells in vitro and in vivo [48]. The utilization of 3D-printed chitosan/alginate scaffolds with nano-HA in cartilage regeneration holds promise in facilitating cell attachment, viability, and exhibiting antibacterial properties [49].

Collagen, a naturally occurring piezoelectric protein (ranging from 0.2 to 2.0 pC/N) stands out as an essential constituent within the ECM of numerous tissues, as previously noted [50]. Its suitability for tissue engineering applications derives from its array of properties, including biocompatibility, robust cell binding capabilities, hydrophilicity, and absorbability within the body [47]. In Ref. [47], increased bone growth within femoral defects in rats upon the implantation of hydrolysis-treated collagen was demonstrated [47]. Additionally, scaffolds containing collagen have proven instrumental in supporting the functionality of SaOS-2 cells [52].

4.2   Piezoceramics

Piezoceramics, widely applied in tissue engineering, owe their popularity to the exceptional combination of their high piezoelectric coefficients and remarkable hardness. These materials generally fall into categories such as lead based, lead free, or titanates. However, limitations exist when considering lead-based ceramics as biomaterials due to their toxicity, even in low doses, narrowing the selection to a few viable options. Among the piezoceramics utilized in bone and cartilage tissue engineering, barium titanate (BT), hydroxyapatite (HA), zinc oxide (ZnO), potassium sodium niobate (KNN), lithium sodium potassium niobate (LKNN), and boron nitride (BN) come forward as important compositions.

BT boasts an impressive piezoelectric property (d_{33} = 191 pC//N), a feature stemming from its perovskite structure that exhibits ferroelectricity below its critical temperature. The inherent displacement between Ti^{4+} and O^{-2} under external force prompts spontaneous polarization even in the absence of an electric field. BT stands as one of the most extensively studied piezoceramics, demonstrating robust biocompatibility in both in vitro and in vivo evaluations. Its innate osteoconductive and osteoinductive qualities stimulate cell proliferation, foster cell attachment, and spur bone growth. Primarily, BT is amalgamated with other bioceramics or polymers to form composites that augment bioactivity and osteogenic potential. For instance, reports indicate that an HA/BT composite exhibits excellent biocompatibility and enhanced bone-forming capabilities [52]. The electrospun PHBV/BT piezoelectric nanohybrid, designed to replicate the natural structure of cartilage, notably facilitated the attachment, proliferation, and expression of the collagen II gene in hMSC-derived chondrocytes, presenting a promising impact on cartilage regeneration [43].

Various forms of HA find application in bone regeneration due to their resemblance to the apatite in bone structure [53]. The amalgamation of HA with other materials aims at optimizing efficiency. For instance, the use of HA filled with osteoinductive and piezoelectric P(VDF/TrFE) nanofibers was shown to impact mesenchymal stem/stromal cells, revealing the potential of HA composites as a favorable agent for enhancing bone regeneration [54].

Zinc plays a pivotal role in regulating cell proliferation and differentiation by modulating the activities of numerous enzymes. When used in the form of piezoelectric ZnO at nanoscale dimensions, it triggers the generation of reactive oxygen species, presenting a potentially toxic effect. However, this cytotoxicity can be mitigated through chemical and physical modifications essential for medical applications. A study recognizing its robust potential in bone tissue regeneration and orthopedic applications demonstrated that the piezoelectric nanocomposite ZnO/PVDF augmented cell density in vitro studies and exhibited antimicrobial properties derived from ZnO [55]. Additionally, Mirza et al. reported that zinc oxide nanoparticles dispersed within a polymeric scaffold in a hypoxic environment displayed the capability to synthesize cartilage [56].

KNN is another well-known piezoceramic (63 pC/N) with biocompatibility, antibacterial, and thermal stability features. KNN is suitable for use in drug delivery systems and tissue (bone, cartilage, skin, and nerve) repair and regeneration due to its piezoelectric feature. Lithium-incorporated KNN (LKNN) exhibits higher piezoelectric response (98 pC/N) and a better chemical stability. However, in the bio-environment, the release profile of Li ions leads to an increase in cytotoxicity. Lithium niobate crystals support bone regeneration by increasing the proliferation of osteoblast cells and osteogenic activity [18].

BNNTs represent a piezoceramic endowed with a substantial piezoelectric coefficient, robust mechanical properties, and a higher surface area [28]. Functioning akin to a nano-vector, this material is readily internalized by cells and delivers electrical stimulation precisely to designated sites to modulate cellular behavior. Proper functionalization involving glycol, chitosan, or surfactants can improve its dispersibility, enhancing cytocompatibility and bolstering its potential for biomedical applications [18]. It was reported that polymeric scaffolds crafted with BNNTs bolster osteoblast proliferation and facilitate attachment and differentiation. However, a deleterious effect on chondrocytes, fibroblasts, and smooth muscle cells was also observed [57]. The boron compounds significantly influence factors like Bone Morphogenetic Proteins (BMP), osteocalcin, RunX2, and mRNA expression, crucial in regulating osteogenic metabolism and consequently modulating the osteoblastic function of cells such as MC3T3-E1 and hMSC [27].

References

1. Xu Q, Gao X, Zhao S, Liu YN, Zhang D, Zhou K, Bowen C et al (2021) Construction of bio-piezoelectric platforms: from structures and synthesis to applications. Adv Mater 33(27):2008452
2. Fang J, Lin T (2019) Energy harvesting properties of electrospun nanofibers. IOP Publishing
3. Pereira, J. P. N., Costa, P., & Lanceros-Méndez, S. (2018). Piezoelectric energy production.
4. Surmenev RA, Orlova T, Chernozem RV, Ivanova AA, Bartasyte A, Mathur S, Surmeneva MA (2019) Hybrid lead-free polymer-based nanocomposites with improved piezoelectric response for biomedical energy-harvesting applications: a review. Nano Energy 62:475–506
5. Carter A, Popowski K, Cheng K, Greenbaum A, Ligler FS, Moatti A (2021) Enhancement of bone regeneration through the converse piezoelectric effect, a novel approach for applying mechanical stimulation. Bioelectricity 3(4):255–271
6. Martin AJP (1941) Tribo-electricity in wool and hair. Proc Phys Soc 53(2):186
7. Kamel NA (2022) Bio-piezoelectricity: fundamentals and applications in tissue engineering and regenerative medicine. Biophys Rev 14(3):717–733
8. Kapat K, Shubhra QT, Zhou M, Leeuwenburgh S (2020) Piezoelectric nano-biomaterials for biomedicine and tissue regeneration. Adv Func Mater 30(44):1909045
9. Guerin S, Tofail SA, Thompson D (2019) Organic piezoelectric materials: milestones and potential. NPG Asia Mater 11(1):10
10. Liu Z, Wan X, Wang ZL, Li L (2021) Electroactive biomaterials and systems for cell fate determination and tissue regeneration: design and applications. Adv Mater 33(32):2007429
11. Kim D, Han SA, Kim JH, Lee JH, Kim SW, Lee SW (2020) Biomolecular piezoelectric materials: from amino acids to living tissues. Adv Mater 32(14):1906989
12. D'Alessandro D, Ricci C, Milazzo M, Strangis G, Forli F, Buda G, Parchi P et al (2021) Piezoelectric signals in vascularized bone regeneration. Biomolecules 11(11):1731
13. Heng BC, Bai Y, Li X, Meng Y, Lu Y, Zhang X, Deng X (2023) The bioelectrical properties of bone tissue. Anim Model Exp Med 6(2):120–130

14. Gracey E, Burssens A, Cambre I, Schett G, Lories R, McInnes IB, Elewaut D et al (2020) Tendon and ligament mechanical loading in the pathogenesis of inflammatory arthritis. Nat Rev Rheumatol 16(4):193–207
15. Popov C, Burggraf M, Kreja L, Ignatius A, Schieker M, Docheva D (2015) Mechanical stimulation of human tendon stem/progenitor cells results in upregulation of matrix proteins, integrins and MMPs, and activation of p38 and ERK1/2 kinases. BMC Mol Biol 16(1):1–11
16. Fernandez-Yague MA, Trotier A, Demir S, Abbah SA, Larrañaga A, Thirumaran A, Biggs MJ et al (2021) A self-powered piezo-bioelectric device regulates tendon repair-associated signaling pathways through modulation of mechanosensitive ion channels. Adv Mater 33(40):2008788
17. Goonoo N, Bhaw-Luximon A (2022) Piezoelectric polymeric scaffold materials as biomechanical cellular stimuli to enhance tissue regeneration. Mater Today Commun 31:103491
18. Jacob J, More N, Kalia K, Kapusetti G (2018) Piezoelectric smart biomaterials for bone and cartilage tissue engineering. Inflamm Regen 38(1):2
19. Leppik L, Oliveira KMC, Bhavsar MB, Barker JH (2020) Electrical stimulation in bone tissue engineering treatments. Eur J Trauma Emerg Surg 46(2):231–244
20. Przekora A (2019) Current trends in fabrication of biomaterials for bone and cartilage regeneration: materials modifications and biophysical stimulations. Int J Mol Sci 20(2):435
21. Sun Y, Liu WZ, Liu T, Feng X, Yang N, Zhou HF (2015) Signaling pathway of MAPK/ERK in cell proliferation, differentiation, migration, senescence and apoptosis. J Recept Signal Transduct 35(6):600–604
22. Thrivikraman G, Boda SK, Basu B (2018) Unraveling the mechanistic effects of electric field stimulation towards directing stem cell fate and function: a tissue engineering perspective. Biomaterials 150:60–86
23. Zhang K, Wang S, Zhou C, Cheng L, Gao X, Xie X, Xu HH et al (2018) Advanced smart biomaterials and constructs for hard tissue engineering and regeneration. Bone Res 6(1):31
24. Ribeiro C, Sencadas V, Correia DM, Lanceros-Méndez S (2015) Piezoelectric polymers as biomaterials for tissue engineering applications. Colloids Surf B: Biointerfaces 136:46–55
25. Tandon B, Blaker JJ, Cartmell SH (2018) Piezoelectric materials as stimulatory biomedical materials and scaffolds for bone repair. Acta Biomater 73:1–20
26. Min S, Lee T, Lee SH, Hong J (2018) Theoretical study of the effect of piezoelectric bone matrix on transient fluid flow in the osteonal lacunocanaliculae. J Orthop Res 36(8):2239–2249
27. Khare D, Basu B, Dubey AK (2020) Electrical stimulation and piezoelectric biomaterials for bone tissue engineering applications. Biomaterials 258:120280
28. Barbosa F, Ferreira FC, Silva JC (2022) Piezoelectric electrospun fibrous scaffolds for bone, articular cartilage and osteochondral tissue engineering. Int J Mol Sci 23(6):2907
29. Zheng T, Huang Y, Zhang X, Cai Q, Deng X, Yang X (2020) Mimicking the electrophysiological microenvironment of bone tissue using electroactive materials to promote its regeneration. J Mater Chem B 8(45):10221–10256
30. Ribeiro C, Pärssinen J, Sencadas V, Correia V, Miettinen S, Hytönen VP, Lanceros-Méndez S (2015) Dynamic piezoelectric stimulation enhances osteogenic differentiation of human adipose stem cells. J Biomed Mater Res Part A 103(6):2172–2175
31. Fukada E (1998) New piezoelectric polymers. Jpn J Appl Phys 37(5S):2775
32. Ohigashi H, Koga K, Suzuki M, Nakanishi T, Kimura K, Hashimoto N (1984) Piezoelectric and ferroelectric properties of P (VDF-TrFE) copolymers and their application to ultrasonic transducers. Ferroelectrics 60(1):263–276
33. Kao FC, Chiu PY, Tsai TT, Lin ZH (2019) The application of nanogenerators and piezoelectricity in osteogenesis. Sci Technol Adv Mater 20(1):1103–1117
34. Damaraju SM, Shen Y, Elele E, Khusid B, Eshghinejad A, Li J, Arinzeh TL et al (2017) Three-dimensional piezoelectric fibrous scaffolds selectively promote mesenchymal stem cell differentiation. Biomaterials 149:51–62
35. Krishnan Y, Grodzinsky AJ (2018) Cartilage diseases. Matrix Biol 71:51–69
36. Xie M, Wang L, Guo B, Wang Z, Chen YE, Ma PX (2015) Ductile electroactive biodegradable hyperbranched polylactide copolymers enhancing myoblast differentiation. Biomaterials 71:158–167

37. Ganeson K, Tan Xue May C, Abdullah AAA, Ramakrishna S, Vigneswari S (2023) Advantages and prospective implications of smart materials in tissue engineering: piezoelectric, shape memory, and hydrogels. Pharmaceutics 15(9):2356
38. Chen Y, Mak AF, Wang M, Li J, Wong MS (2006) PLLA scaffolds with biomimetic apatite coating and biomimetic apatite/collagen composite coating to enhance osteoblast-like cells attachment and activity. Surf Coat Technol 201(3–4):575–580
39. Prabhakaran MP, Venugopal J, Ramakrishna S (2009) Electrospun nanostructured scaffolds for bone tissue engineering. Acta Biomater 5(8):2884–2893
40. Ko EK, Jeong SI, Rim NG, Lee YM, Shin H, Lee BK (2008) In vitro osteogenic differentiation of human mesenchymal stem cells and in vivo bone formation in composite nanofiber meshes. Tissue Eng Part A 14(12):2105–2119
41. Zhou Z, Zheng J, Meng X, Wang F (2023) Effects of electrical stimulation on articular cartilage regeneration with a focus on piezoelectric biomaterials for articular cartilage tissue repair and engineering. Int J Mol Sci 24(3):1836
42. Gorodzha SN, Muslimov AR, Syromotina DS, Timin AS, Tcvetkov NY, Lepik KV, Surmenev RA et al (2017) A comparison study between electrospun polycaprolactone and piezoelectric poly (3-hydroxybutyrate-co-3-hydroxyvalerate) scaffolds for bone tissue engineering. Colloids Surf B: Biointerfaces 160:48–59
43. Jacob J, More N, Mounika C, Gondaliya P, Kalia K, Kapusetti G (2019) Smart piezoelectric nanohybrid of poly (3-hydroxybutyrate-co-3-hydroxyvalerate) and barium titanate for stimulated cartilage regeneration. ACS Appl Bio Mater 2(11):4922–4931
44. Zaborowska M, Bodin A, Bäckdahl H, Popp J, Goldstein A, Gatenholm P (2010) Microporous bacterial cellulose as a potential scaffold for bone regeneration. Acta Biomater 6(7):2540–2547
45. Dong S, Zhang Y, Mei Y, Zhang Y, Hao Y, Liang B, Niu L et al (2022) Researching progress on bio-reactive electrogenic materials with electrophysiological activity for enhanced bone regeneration. Front Bioeng Biotechnol 10:921284
46. Li Y, Xun X, Xu Y, Zhan A, Gao E, Yu F, Yang C et al (2022) Hierarchical porous bacterial cellulose scaffolds with natural biomimetic nanofibrous structure and a cartilage tissue-specific microenvironment for cartilage regeneration and repair. Carbohyd Polym 276:118790
47. Puppi D, Chiellini F, Piras AM, Chiellini E (2010) Polymeric materials for bone and cartilage repair. Prog Polym Sci 35(4):403–440
48. Chen YH, Tai HY, Fu E, Don TM (2019) Guided bone regeneration activity of different calcium phosphate/chitosan hybrid membranes. Int J Biol Macromol 126:159–169
49. Sadeghianmaryan A, Naghieh S, Yazdanpanah Z, Sardroud HA, Sharma NK, Wilson LD, Chen X (2022) Fabrication of chitosan/alginate/hydroxyapatite hybrid scaffolds using 3D printing and impregnating techniques for potential cartilage regeneration. Int J Biol Macromol 204:62–75
50. Ferreira AM, Gentile P, Chiono V, Ciardelli G (2012) Collagen for bone tissue regeneration. Acta Biomater 8(9):3191–3200
51. Elkasabgy NA, Mahmoud AA, Shamma RN (2018) Determination of cytocompatibility and osteogenesis properties of in situ forming collagen-based scaffolds loaded with bone synthesizing drug for bone tissue engineering. Int J Polym Mater Polym Biomater 67(8):494–500
52. Tang Y, Wu C, Wu Z, Hu L, Zhang W, Zhao K (2017) Fabrication and in vitro biological properties of piezoelectric bioceramics for bone regeneration. Sci Rep 7(1):43360
53. Lin K, Wu C, Chang J (2014) Advances in synthesis of calcium phosphate crystals with controlled size and shape. Acta Biomater 10(10):4071–4102
54. Barbosa F, Garrudo FF, Alberte PS, Resina L, Carvalho MS, Jain A, Silva JC et al (2023) Hydroxyapatite-filled osteoinductive and piezoelectric nanofibers for bone tissue engineering. Sci Technol Adv Mater 24(1):2242242
55. Li Y, Sun L, Webster TJ (2018) The investigation of ZnO/poly (vinylidene fluoride) nanocomposites with improved mechanical, piezoelectric, and antimicrobial properties for orthopedic applications. J Biomed Nanotechnol 14(3):536–545
56. Mirza EH, Pan-Pan C, Wan Ibrahim WMAB, Djordjevic I, Pingguan-Murphy B (2015) Chondroprotective effect of zinc oxide nanoparticles in conjunction with hypoxia on bovine cartilage-matrix synthesis. J Biomed Mater Res Part A 103(11):3554–3563

57. Ciofani G, Danti S, Genchi GG, Mazzolai B, Mattoli V (2013) Boron nitride nanotubes: biocompatibility and potential spill-over in nanomedicine. Small 9(9–10):1672–1685

Artificial Intelligence in Predicting Hard Tissue Regeneration: Current Situation and Upcoming Perspectives

Nura Brimo, Dilek Çökeliler Serdaroğlu, Halit Muhittin, Mustafa Kaplan, and Abdulwahab Omira

Abstract This chapter provides an overview of the role of Artificial Intelligence (AI) in the field of regenerative medicine, specifically focusing on its application in tissue regeneration, with a particular emphasis on hard tissue regeneration. The chapter opens with an introduction to AI, highlighting its relevance in regenerative medicine. It then progresses to discuss diverse AI techniques employed for predicting tissue regeneration. Machine learning algorithms, including deep learning networks, have shown promise in accurately predicting the regenerative potential of different tissues, which have shown promising results in accurately predicting the regenerative possibility of various tissues. The possibility of AI to revolutionize the field of hard tissue regeneration is then explored. The chapter further delves into the specific applications of AI in hard tissue regeneration, with a focus on dental and orthopedic applications. In the dental field, AI has shown promise in improving clinical decision-making, optimizing dental implant placement, and predicting the success of bone grafting procedures. Similarly, in the orthopedic field, AI has facilitated the development of personalized treatment plans, improved surgical outcomes, and enhanced fracture healing. Despite these challenges and limitations associated with AI, the chapter highlights the potential of AI to overcome these limitations and further drive advancements in the field of regenerative medicine.

N. Brimo · D. Ç. Serdaroğlu (✉)
Department of Biomedical Engineering, Baskent University, Baglıca Campus, 06530 Ankara, Turkey
e-mail: cokeliler@baskent.edu.tr

N. Brimo
Innovation, Research, and Development, The White Helmets Organization/Les Casques Blancs, Montreal 63005, Canada

H. Muhittin
Faculty of Biology, Medicine and Health, University of Manchester, Oxford, Manchester M13 9PL, United Kingdom

M. Kaplan
Bioinformatics and Systems Biology, University of Manchester, Manchester, United Kingdom

A. Omira
Computer Science Department, Stanford University, 06530, Stanford, CA 94305, USA

Keywords Artificial intelligence · Regeneration · Hard tissue · Machine learning

1 Introduction

Tissue regeneration is a complex and critical process in the field of medicine, holding the possibility to restore function and improve the quality of life for patients suffering from various injuries and diseases. However, accurately predicting tissue regeneration and understanding its underlying mechanisms remain significant challenges. Hard tissue regeneration is particularly challenging due to the complexity of the interactions between cells, extracellular matrix components, and signaling molecules. Accurately predicting tissue regeneration outcomes, such as the success of bone graft procedures or the healing of chronic wounds, remains a significant challenge. With the advancements in science and technology, the integration of artificial intelligence (AI) has emerged as a promising approach to deciphering the intricacies of tissue regeneration and developing predictive models for its success. Tissue regeneration encompasses the natural healing process of damaged or lost tissues in living organisms. It involves the replacement or repair of injured or diseased tissue to restore normal structure and function [1, 2]. From simple cutaneous wounds to complex organ damage, tissue regeneration holds immense potential in the field of regenerative medicine. Predicting tissue regeneration is crucial for advancing therapeutic interventions and enhancing patient outcomes. However, this prediction is a challenging task that necessitates a comprehensive understanding of cellular and molecular events, as well as environmental factors that influence tissue regeneration. Traditional approaches have relied on empirical data and experimentation, leading to slow progress and limited accuracy in predicting outcomes [3, 4].

Hard tissue regeneration, such as bone and tooth regeneration, is an intricate process including complex interactions between cells, extracellular matrix components, and signaling molecules. The ability to accurately predict the outcomes of regenerative processes in hard tissues has significant implications for personalized medicine and the development of novel therapeutic strategies. Recently, artificial intelligence (AI) has emerged as a promising tool for predicting tissue regeneration outcomes. This chapter aims to provide a comprehensive overview of the current situation and upcoming perspectives regarding the use of AI in predicting hard tissue regeneration. Hard tissue regeneration refers to the process of restoring lost or damaged hard tissues, such as bone and teeth. Traditionally, predicting the success of such regenerative procedures has relied on expert knowledge, empirical evidence, and clinical experience [5–7]. However, these methods often lack precision and can be time-consuming. The integration of AI has the potential to revolutionize this field by providing accurate predictive models for hard tissue regeneration.

Artificial intelligence refers to the development of computer systems able to perform tasks that mainly need human intelligence. AI algorithms can analyze large datasets, identify patterns, and make predictions or decisions based on these patterns. In the context of tissue regeneration, AI has the potential to revolutionize our understanding by providing insights into complex molecular interactions and predicting tissue regeneration outcomes. AI algorithms can efficiently mine vast datasets obtained from diverse sources, including genomics, proteomics, and clinical records. By analyzing these datasets, AI can identify novel biomarkers and genetic factors associated with successful tissue regeneration. Furthermore, AI can help in uncovering previously unrecognized relationships between various biological components, paving the way for new predictive models [8, 9]. In this chapter, we will find the current state of research in the field of artificial intelligence for predicting hard tissue regeneration outcomes. Additionally, we will discuss the potential future directions and upcoming perspectives that hold promise for advancing this field [10–14].

2 Introduction to Artificial Intelligence (AI) and Machine Learning Algorithms

In the late 20th century, the development of algorithms capable of building predictive models for systematic data analysis marked a pivotal moment in technology. These algorithms, later termed machine learning (ML) algorithms due to their ability to emulate the way humans learn from experience, became a cornerstone of data-driven research methodologies. As the 21st century approached, advancements in parallel computing and data storage transformed the potential of these algorithms into practical solutions for complex problems. By the 1990s, the integration of intelligent systems and ML algorithms gained significant traction, finding applications across a wide range of fields.

2.1 Development of Artificial Intelligence and Machine Learning Methodologies

There has been critical progress in the advancement of AI and ML methodologies. AI includes the creation of intelligent machines capable of performing tasks that systematically necessitate human intelligence, such as visual perception, speech recognition, and decision-making. On the other hand, ML is a subset of AI that concentrates on algorithms and statistical models, allowing computers to learn and make predictions or decisions without explicit programming.

One of the earliest milestones in AI and ML was the development of the perceptron algorithm by Frank Rosenblatt in the late 1950s. This algorithm formed the basis for neural networks, which are layered networks of interconnected nodes that mimic the structure and function of the human brain. Neural networks have been successfully implemented in many aspects, including image and speech recognition, natural language processing, and autonomous driving. Another important development in

ML was the introduction of support vector machines (SVMs) by Vladimir Vapnik and his colleagues in 1990s. SVMs are supervised learning models that are used for classification and regression analysis. They have been widely used in areas such as bioinformatics, computer vision, and finance [15, 16].

The field of deep learning, which involves training neural networks with multiple hidden layers, has also seen significant advancements in recent years. In 2012, a deep learning algorithm called AlexNet achieved a breakthrough in image recognition by significantly outperforming other methods in the ImageNet Large Scale Visual Recognition Challenge. Since then, deep learning has become a state-of-the-art technique in domains such as computer vision and natural language processing. The development of AI and ML methodologies has been supported by advancements in computing power and the availability of large datasets. The advent of graphical processing units (GPUs) has enabled researchers to train complex neural networks more efficiently. In addition, the rise of big data has provided researchers with vast amounts of data to train and validate their models. Furthermore, the open-source community has played a crucial role in the development of AI and ML methodologies. Frameworks such as TensorFlow, PyTorch, and scikit-learn have made it easier for researchers and developers to implement and experiment with various algorithms and models [17–19].

2.2 Branches of Artificial Intelligence (AI)

The various branches of Artificial Intelligence (AI) have revolutionized our interaction with technology, driving innovation across multiple fields. This overview highlights key areas within AI, such as natural language processing (NLP), machine learning (ML), supervised learning, and unsupervised learning.

A critical branch of AI, Natural Language Processing (NLP), enables machines to understand, interpret, and generate human language. NLP encompasses subfields like Natural Language Understanding (NLU) and Natural Language Generation (NLG), using algorithms and statistical models to process and respond to human communication. Applications of NLP include virtual assistants, language translation tools, sentiment analysis, chatbots, and more.

Machine Learning (ML) (see Fig. 1), another foundational area of AI, allows systems to learn and improve from data without explicit programming. ML techniques are categorized primarily into supervised learning and unsupervised learning. In supervised learning, models are trained on labeled datasets, where inputs are paired with known outputs. This approach enables predictions or classifications for new, unseen data and is widely applied in tasks such as image recognition, text classification, and fraud detection. Algorithms commonly used in supervised learning include linear regression, decision trees, support vector machines, and neural networks. Unsupervised learning, on the other hand, involves analyzing and identifying patterns in unlabeled data, often applied in clustering and dimensionality

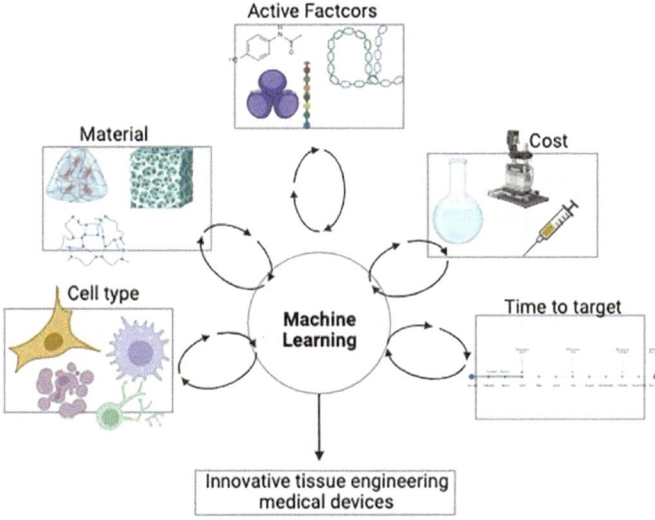

Fig. 1 Using machine learning to design unique tissue engineering medical devices

reduction tasks. Together, these branches of AI continue to shape advancements in technology, creating smarter and more adaptive systems [20, 21].

Otherwise, unsupervised learning deals with unlabeled data, where the algorithm tends to find patterns or structures in the data without any predefined output labels. Clustering and dimensionality reduction are common tasks in unsupervised learning. Clustering algorithms group similar data points together while dimensionality reduction techniques aim to reduce the number of features or variables in the data. Unsupervised learning has applications in recommendation systems, anomaly detection, and data visualization [22].

3 Artificial Intelligence (AI) in Regenerative Medicine

Artificial Intelligence (AI) techniques have been extensively employed in regenerative medicine to enhance diagnosis, treatment planning, and tissue engineering processes. The integration of AI with regenerative medicine has shown promising results in improving patient outcomes and accelerating the development of regenerative therapies. In this response, we will discuss some of the key AI techniques that have been applied in regenerative medicine, along with relevant citations and references.

ML algorithms have been widely applied in analyzing huge datasets of patient information, genetic data, and biomaterial properties to identify patterns and make predictions. ML techniques like decision trees, random forests, and support vector machines have been utilized for the development of predictive models for disease

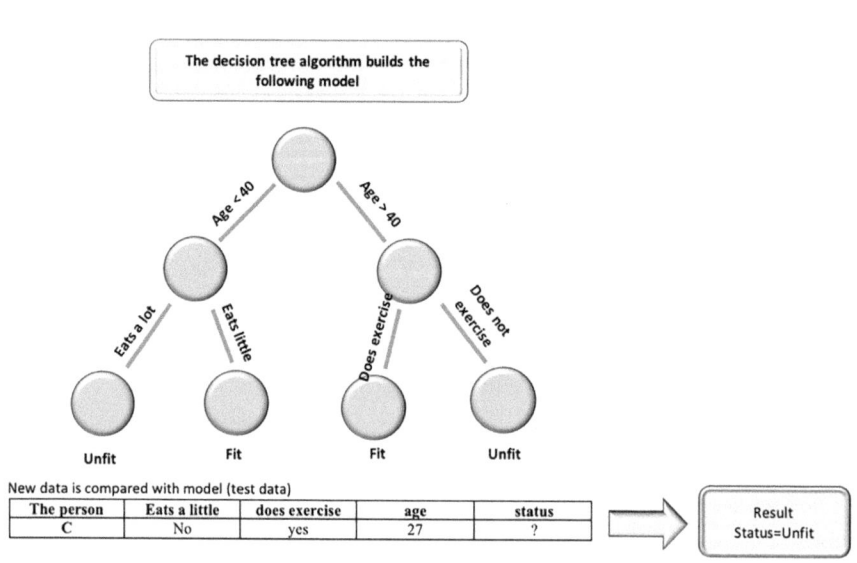

(Available Data)

The person	Eats a little	does exercise	age	status
X	Yes	No	35	Fit
Y	No	No	44	Unfit

The decision tree algorithm builds the following model

Age < 40   Age > 40

Eats a lot   Eats little   Does exercise   Does not exercise

Unfit   Fit   Fit   Unfit

New data is compared with model (test data)

The person	Eats a little	does exercise	age	status
C	No	yes	27	?

Result
Status=Unfit

Fig.2 Decision tree (interpretable)

progression, treatment response, and tissue regeneration outcomes [23]. One example of the successful implementation of ML in regenerative medicine is the prediction of the differentiation potential of stem cells based on their gene expression profiles [24]. This study used ML algorithms to identify gene expression patterns associated with differentiating stem cells into specific lineages, enabling accurate predictions of cell fate (see Fig. 2).

Deep Learning (DL) has gained significant attention in regenerative medicine due to its ability to learn complex patterns and extract features from large and diverse datasets. Convolutional Neural Networks (CNNs) and Recurrent Neural Networks (RNNs) are similar DL techniques employed for image analysis, genetic sequence analysis, and natural language processing in the context of regenerative medicine. For example, CNNs have been used for automated analysis of histological images to classify and quantify different types of tissue structures, helping in the diagnosis and assessment of regenerative therapies [25]. DL techniques have also been utilized for drug discovery, where RNNs have been employed to generate new molecular structures with regenerative potential [3].

The integration of multi-modal data, such as genomic, proteomic, imaging, and clinical data, is crucial in regenerative medicine to identify biomarkers, understand disease mechanisms, and develop personalized treatment strategies. AI techniques, including Bayesian networks, graphical models, and knowledge-based reasoning, are employed for data fusion and integration. For instance, the Genomic Data Integration (GDI) framework combines gene expression data with protein–protein interaction networks to identify key genes and signaling pathways involved in tissue regeneration. GDI enables the identification of potential therapeutic targets for regenerative medicine.

4 Artificial Intelligence (AI) Techniques for Predicting Tissue Regeneration

Tissue regeneration plays a pivotal role in the field of regenerative medicine, aiming to restore damaged or degenerated tissues and organs. The ability to predict tissue regeneration outcomes is essential in developing effective therapeutic strategies and personalized treatment approaches. In recent years, the integration of artificial intelligence (AI) techniques with tissue regeneration research has shown great promise in improving prediction accuracy. This article explores the various AI techniques utilized for predicting tissue regeneration and provides relevant citations and references for further reading.

Machine learning algorithms have been employed extensively in tissue regeneration research due to their ability to analyze large datasets and identify complex patterns. Some commonly used algorithms include Random Forest: Random Forest algorithms have been utilized to predict tissue regeneration outcomes based on various variables, such as patient demographics, injury types, and treatment protocols [4]. Support Vector Machines (SVM): SVM algorithms have demonstrated efficacy in predicting tissue regeneration by classifying different types of tissue samples based on gene expression patterns [26].

Deep learning techniques have gained attention in tissue regeneration research due to their ability to automatically learn hierarchies of features from large datasets. Some key applications include Convolutional Neural Networks (CNN): CNNs have been widely employed in predicting tissue regeneration outcomes by analyzing medical imaging data, such as histopathological images or magnetic resonance imaging (MRI) scans [27]. Recurrent Neural Networks (RNN): RNNs have been utilized to predict tissue regeneration kinetics by analyzing temporal data, such as gene expression profiles or dynamic imaging sequences [28].

AI techniques for tissue regeneration prediction often incorporate data from diverse sources to improve accuracy. Some notable approaches include Multimodal Data Fusion: Combining data from multiple sources, such as clinical data, genetic information, and imaging data, can enhance tissue regeneration prediction models [29]. Transfer Learning: Transfer learning techniques have been used to leverage pretrained AI models from other domains (e.g., image recognition or natural language processing) to enhance tissue regeneration prediction accuracy [9, 30].

Artificial intelligence (AI) technologies, such as machine learning (ML) algorithms, deep learning techniques, and data fusion/integration methods, have demonstrated remarkable potential in predicting outcomes in tissue regeneration. These tools empower researchers to analyze intricate interactions among biological, environmental, and clinical factors, enhancing the development of optimized treatment strategies in regenerative medicine. As AI continues to evolve, ongoing advancements in this field are expected to lead to more precise predictions and personalized approaches, revolutionizing the way tissue regeneration is understood and applied in healthcare.

5 Machine Learning (ML) and Artificial Intelligence (AI) Approaches for Tissue Scaffolds Design

Machine Learning (ML) and Artificial Intelligence (AI) are increasingly being applied as powerful tools in various fields, including tissue engineering. Over the past few years, there has been a rising interest in applying AI and ML approaches to the design and production of tissue scaffolds, which play a crucial role in tissue regeneration and repair. One particular area where AI has made significant contributions is in AI-aided design strategies for tissue engineering scaffolds. Traditionally, scaffold design heavily relies on trial-and-error methods, which can be both time-consuming and inefficient. However, with the use of AI algorithms, large datasets can be analyzed to identify patterns and relationships between scaffold properties and desired outcomes. This allows for the development of optimized scaffold designs. By leveraging AI, researchers have been able to optimize the mechanical properties of scaffolds to enhance cell adhesion and proliferation [31].

ML-assisted 2D and 3D printing of tissue engineering scaffolds is another area of active research. ML algorithms can analyze various parameters involved in the printing process, such as printing speed, nozzle diameter, and material composition, to optimize the printing parameters for desired scaffold properties. For instance, ML has been used to control the printing process and improve the precision of scaffold fabrication, enabling the creation of complex geometries with high accuracy and reproducibility [32].

Furthermore, AI and ML have been employed in the field of in vitro imaging of cells on scaffolds. By analyzing large volumes of imaging data, AI algorithms can automatically detect and quantify cell morphology, proliferation, and differentiation. This allows for the assessment of the biocompatibility and functionality of the scaffolds with minimal human intervention. Such automated analysis can save time and improve the objectivity and consistency of the evaluation results [33].

Looking ahead, the integration of artificial intelligence (AI) and machine learning (ML) with tissue scaffold manufacturing holds tremendous potential. One promising application is the creation of AI-driven biofabrication systems capable of designing scaffolds with customized properties tailored to individual patient needs. For instance, AI algorithms can process patient-specific medical data, such as CT or MRI scans, to generate scaffold designs optimized to match the unique anatomical structure and requirements of the targeted tissue or organ. This approach could significantly enhance the precision and effectiveness of regenerative therapies.

In addition, AI and ML can be utilized in the domain of scaffold material optimization. ML algorithms can evaluate the performance of different biomaterial combinations based on various criteria, such as biocompatibility, mechanical properties, and degradation rate. By considering a wide range of materials and their interactions, AI can help identify novel materials or combinations that exhibit superior properties for tissue scaffold applications [34, 35].

In summary, AI and ML approaches hold great promise for advancing tissue scaffold design and manufacturing. These techniques enable the development of optimized scaffold designs, precise scaffold fabrication, automated evaluation of scaffold–cell interactions, and the integration of patient-specific requirements into scaffold manufacturing processes. Continued research in this field will undoubtedly lead to further advancements in tissue engineering and regenerative medicine.

6 Artificial Intelligence (AI) and Its Applications in Hard Tissue Regeneration

AI has shown great potential in revolutionizing different fields of healthcare, including hard tissue regeneration. Hard tissues, such as bones and teeth, play a crucial role in providing structural support to the human body. When these tissues become damaged due to trauma, disease, or aging, they require regeneration for optimal functionality and restoration. AI, specifically ML algorithms, has been widely applied in the field of hard tissue regeneration to enhance the diagnosis, treatment, and monitoring processes. One of the key applications of AI is in the area of medical imaging. It enables the accurate and efficient analysis of radiographic images, such as X-rays and CT scans, for the identification and characterization of hard tissue defects. AI algorithms can aid in early detection, precise classification, and quantification of bone disorders, facilitating personalized treatment planning and monitoring. Furthermore, AI has found utility in the development of biomaterials and scaffolds for tissue engineering. By utilizing machine learning algorithms, researchers can efficiently design and screen a vast number of potential biomaterial compositions and scaffold structures. This accelerates the discovery and optimization of biomaterials that promote hard tissue regeneration, ultimately leading to improved treatment outcomes [36–41]. Supervised learning algorithms can be used to identify patterns in patient data that can predict the success of tissue regeneration procedures. Unsupervised learning algorithms can be used to identify novel biomarkers and genetic factors associated with successful tissue regeneration. Reinforcement learning algorithms can be used to optimize the design of biomaterials and scaffolds that promote tissue regeneration.

Apart from tissue engineering, AI has also demonstrated its potential in guiding surgical interventions for hard tissue regeneration. By integrating AI algorithms into surgical systems, surgeons can benefit from real-time assistance during complex procedures. These algorithms can help in identifying critical anatomical structures, guiding the surgeon's movements, and providing precise feedback on surgical progress. This enhances surgical accuracy and reduces the risk of complications during tissue regeneration surgeries. While AI has made significant strides in hard tissue regeneration, it is important to acknowledge the challenges and limitations associated with its implementation. One of the key concerns is the lack of robust and diverse datasets required for training AI algorithms. It is crucial to gather comprehensive and representative data to ensure accurate and unbiased predictions. Additionally, ethical considerations surrounding data privacy, patient consent, and algorithm transparency need to be carefully addressed to ensure the responsible use of AI in healthcare [42–44]. In conclusion, AI has emerged as a powerful tool in hard tissue regeneration, providing researchers and clinicians with valuable insights, enhanced diagnostics, and improved treatment outcomes. The integration of AI algorithms into medical imaging, biomaterial development, and surgical interventions has the potential to significantly enhance the field and pave the way for innovative and personalized regenerative therapies.

7 Challenges and Limitations

Despite the promising applications of AI in hard tissue regeneration, several challenges and limitations need to be considered. Firstly, the scarcity and heterogeneity of high-quality data pose significant obstacles to developing accurate predictive models. Large-scale collaborative efforts involving multiple research institutions and data-sharing initiatives are essential to address these challenges.

While the potential of artificial intelligence (AI) in hard tissue regeneration is undeniable, several challenges and limitations must be addressed for its effective implementation. One major hurdle is the scarcity and variability of high-quality data, which hampers the development of reliable predictive models. Collaborative efforts across research institutions and data-sharing initiatives are essential to overcome this obstacle and create diverse, robust datasets for training AI systems.

Another critical issue is data privacy and security. AI systems rely on extensive datasets for training and validation, raising concerns about the exposure of sensitive patient information. To mitigate this risk, stringent data protection protocols and anonymization techniques must be employed to safeguard patient privacy while enabling the use of data for research.

The complexity of modeling biological systems presents an additional challenge. Biological tissues are inherently dynamic and multifaceted, making it difficult for AI models to accurately simulate and predict tissue behavior under diverse conditions. This calls for the development of more advanced AI algorithms capable of capturing the intricate nature of biological processes.

Interpretability of AI models also remains a significant concern. In regenerative medicine, understanding the underlying biological mechanisms is crucial for designing effective therapies. However, many AI models, especially deep learning algorithms, operate as black boxes, offering accurate predictions without explaining the processes behind them. To address this, researchers should prioritize the development of explainable AI models that provide both accurate predictions and valuable biological insights.

Finally, ethical considerations surrounding AI in tissue regeneration must not be overlooked. Issues such as patient privacy, data security, and potential biases in algorithms need careful attention to ensure the responsible and fair application of AI in healthcare. By addressing these challenges, the integration of AI in regenerative medicine can progress responsibly, unlocking its full potential to revolutionize patient care.

8   Upcoming Perspectives and Future Prospects

The future of AI in predicting hard tissue regeneration is ripe with possibilities, including the exploration of AI-driven biomaterial design and the integration of AI in post-surgical monitoring. Advancements in technologies, such as single-cell sequencing, mass spectrometry, and three-dimensional tissue imaging, will provide detailed molecular and cellular profiles, leading to more accurate predictive models. Integration of data from multiple sources, including clinical data, will enable the development of personalized predictive models for regenerative therapies. Efforts should be made to enhance the availability and quality of data for tissue regeneration research. This can include collaborative initiatives to collect and share datasets, establishment of standardized protocols for data collection, and development of techniques for non-invasive monitoring of tissue regeneration processes [3, 4, 45]. By improving data accessibility and quality, researchers would have more robust foundations for developing predictive models. Future research should focus on developing integrative approaches that capture the complexity of tissue regeneration processes. This can involve the incorporation of multi-omics data, spatial information, and systems biology modeling techniques. By encompassing a broader range of variables and interactions, these integrative approaches have the potential to provide a more accurate understanding of tissue regeneration mechanisms. Explainable AI Models: Given the importance of interpretability and explainability, future research should prioritize the development of AI models that can provide transparent insights into their predictions. Techniques such as interpretable machine learning and model-agnostic explanations can help bridge the gap between AI predictions and biological understanding, enhancing trust and usability in clinical settings [46, 47].

9 Conclusion

In this chapter, we have explored the current situation and upcoming perspectives on using Artificial Intelligence (AI) for predicting hard tissue regeneration. Through a comprehensive analysis of various studies and advancements in the field, we have gained valuable insights into the potential of these technologies in revolutionizing the way we approach tissue regeneration. The main findings of this chapter highlight the significant advancements made in AI and their application in predicting hard tissue regeneration. AI algorithms have proven to be powerful tools in processing and analyzing huge amounts of data, enabling researchers to identify patterns and make accurate predictions regarding tissue regeneration outcomes.

The implications of these findings are profound. The ability to predict tissue regeneration outcomes can greatly improve clinical decision-making and treatment planning. By analyzing patient data and combining it with AI algorithms, physicians can anticipate the regenerative potential of tissues and tailor treatment approaches accordingly. This personalized medicine approach is able to lead to enhanced patient results and decrease the time and resources needed for trial-and-error treatment approaches.

However, it is essential to acknowledge that while AI has shown great promise in tissue regeneration prediction, there are still required challenges to be addressed. One such challenge is the need for high-quality and standardized data. As AI algorithms heavily rely on data inputs, ensuring the availability of accurate and comprehensive datasets is crucial for their successful implementation. Furthermore, ethical considerations regarding patient data privacy and consent have to be carefully addressed to maintain trust and ensure the responsible use of AI in healthcare.

Looking ahead, future research and applications in the area of AI for tissue regeneration prediction should focus on addressing these challenges while further advancing the capabilities of these technologies. Additionally, efforts should be made to establish standardized protocols and guidelines for data collection, sharing, and analysis to promote consistency and comparability among studies.

In conclusion, AI has not only emerged as a powerful tool in predicting hard tissue regeneration outcomes but also stands as a beacon for innovative medical approaches in the future. The findings presented in this chapter underscore the immense potential of these technologies, as well as the need for further research and collaboration to overcome current challenges. With continued advancements and interdisciplinary efforts, the future of tissue regeneration prediction using AI looks incredibly promising.

References

1. Dorozhkin SV (2022) Calcium orthophosphate (CaPO4)-based bioceramics: preparation, properties, and applications. Coatings 12(10):1380
2. Al-Kharusi G et al (2022) The role of machine learning and design of experiments in the advancement of biomaterial and tissue engineering research. Bioengineering 9(10):561

3. Nosrati H, Nosrati M (2023) Artificial intelligence in regenerative medicine: applications and implications. Biomimetics 8(5):442
4. Vl Bitkina O, Park J, Kim HK (2023) Application of artificial intelligence in medical technologies: a systematic review of main trends. Digit Health 9:20552076231189331
5. Adir O et al (2020) Integrating artificial intelligence and nanotechnology for precision cancer medicine. Adv Mater 32(13):1901989
6. Chairat S et al (2023) Ai-assisted assessment of wound tissue with automatic color and measurement calibration on images taken with a smartphone. Healthcare 11(2):273. MDPI
7. Gao W et al (2022) Application of medical imaging methods and artificial intelligence in tissue engineering and organ-on-a-chip. Front Bioeng Biotechnol 10:985692
8. Koromina M, Pandi M-T, Patrinos GP (2019) Rethinking drug repositioning and development with artificial intelligence, machine learning, and omics. Omics: J Integr Biol 11:539–548
9. Javaid S et al (2023) Identification and ranking biomaterials for bone scaffolds using machine learning and PROMETHEE. Res Biomed Eng 39(1):129–138
10. Hunter B, Hindocha S, Lee RW (2022) The role of artificial intelligence in early cancer diagnosis. Cancers 14(6):1524–1524
11. Jones MA et al (2022) Applying artificial intelligence technology to assist with a breast cancer diagnosis and prognosis prediction. Front Oncol 12:980793
12. Kim C et al (2018) Polymer genome: a data-powered polymer informatics platform for property predictions. J Phys Chem C 122(31):17575–17585
13. Kwaria RJ et al (2020) Data-driven prediction of protein adsorption on self-assembled monolayers toward material screening and design. ACS Biomater Sci Eng 6(9):4949–4956
14. Madani M, Behzadi MM, Nabavi S (2022) The role of deep learning in advancing breast cancer detection using different imaging modalities: a systematic review. Cancers 14(21)
15. Rosenblatt F (1958) The perceptron: a probabilistic model for information storage and organization in the brain. Psychol Rev 65(6):386
16. Vapnik V (1999) The nature of statistical learning theory. Springer Science & Business Media
17. Krizhevsky A, Sutskever I, Hinton GE (2012) Imagenet classification with deep convolutional neural networks. Adv Neural Inf Process Syst 25
18. Hardesty L (2016) Making computers explain themselves. MIT News 27:2016
19. Pedregosa F et al (2011) Scikit-learn: machine learning in python. J Mach Learn Res 12:2825–2830
20. Meurers D (2012) Natural language processing and language learning. Encycl Appl Linguist, 4193–4205
21. Bohr A, Memarzadeh K (2020) The rise of artificial intelligence in healthcare applications. In: Artificial intelligence in healthcare. Elsevier, pp 25–60
22. Badai J, Bu Q, Zhang L (2020) Review of artificial intelligence applications and algorithms for brain organoid research. Interdiscip Sci: Comput Life Sci 12:383–394
23. Galat V et al (2012) A model of early human embryonic stem cell differentiation reveals inter- and intracellular changes on transition to squamous epithelium. Stem Cells Dev 21(8):1250–1263
24. Zhang C et al (2019) Improved generative adversarial networks using the total gradient loss for the resolution enhancement of fluorescence images. Biomed Opt Express 10(9):4742–4756
25. G´omez-Bombarelli R et al (2018) Automatic chemical design using a data-driven continuous representation of molecules. ACS Cent Sci 4(2):268–276
26. Sujeeun LY et al (2020) Correlating in vitro performance with physicochemical characteristics of nanofibrous scaffolds for skin tissue engineering using supervised machine learning algorithms". R Soc Open Sci 7(12):201293
27. Furey TS et al (2000) Support vector machine classification and validation of cancer tissue samples using microarray expression data. Bioinformatics 16(10):906–914
28. Xu J et al (2016) A deep convolutional neural network for segmenting and classifying epithelial and stromal regions in histopathological images. Neurocomputing 191:214–223
29. Movahhedi M et al (2023) Predicting 3D soft tissue dynamics from 2D imaging using physics informed neural networks. Commun Biol 6(1):541

30. Kline A et al (2022) Multimodal machine learning in precision health: a scoping review. npj Digit Med 5(1):171
31. Barrera B, Dolores M, Franco-Martínez F, Lantada AD (2021) Artificial intelligence aided design of tissue engineering scaffolds employing virtual tomography and 3D Convolutional Neural Networks. Materials 14(18):5278
32. Rojek I et al (2020) AI-optimized technological aspects of the material used in 3D printing processes for selected medical applications. Materials 13(23):5437
33. Nayak VV et al (2023) 3D printing type 1 bovine collagen scaffolds for tissue engineering applications—physicochemical characterization and in vitro evaluation. Gels 9(8):637
34. Scherr T et al (2020) Cell segmentation and tracking using CNN-based distance predictions and a graph-based matching strategy. PLoS One 12:e0243219
35. Altyar AE et al (2023) Future regenerative medicine developments and their therapeutic applications. Biomed Pharmacother 158:114131
36. Zhong F, Jiang Y (2019) Endogenous pancreatic β cell regeneration: a potential strategy for the recovery of β cell deficiency in diabetes. Front Endocrinol 10:101
37. Ossowska A, Kusiak A, Swietlik D (2022) Artificial intelligence in dentistry—narrative review. Int J Environ Res Public Health 19(6):3449
38. Wong SK et al (2023) A review of the application of natural and synthetic scaffolds in bone regeneration. J Funct Biomater 14(5):286
39. Mackay BS et al (2021) The future of bone regeneration: integrating AI into tissue engineering. Biomed Phys Eng Express 7(5):052002
40. Gupta R et al (2021) Artificial intelligence to deep learning: machine in intelligence approach for drug discovery. Mol Divers 25:1315–1360
41. Ferguson AL, Brown KA (2022) Data-driven design and autonomous experimentation in soft and biological materials engineering. Annu Rev Chem Biomol Eng 13:25–44
42. Thakur A et al (2020) Application of artificial intelligence in pharmaceutical and biomedical studies. Curr Pharm Des 26(29):3569–3578
43. Mak K-K, Pichika MR (2019) Artificial intelligence in drug development: present status and future prospects. Drug Discov Today 24(3):773–780
44. Nsugbe E (2023) An artificial intelligence-based decision support system for early diagnosis of polycystic ovaries syndrome. Healthc Anal 3:100164
45. Najafi H et al (2021) Recent advances in design and applications of biomimetic self-assembled peptide hydrogels for hard tissue regeneration. Bio-Des Manuf 4:735–756
46. Popa L et al (2022) Bacterial cellulose—a remarkable polymer as a source for biomaterials tailoring. Materials 15(3):1054
47. Ortiz-Catalan M et al (2023) A highly integrated bionic hand with neural control and feedback for use in daily life. Sci Robot 8(83):eadf7360